SAVE THAT ENERGY

As costs climb higher and supplies run low, it becomes more important than ever to save energy. This book shows how to check out your home, your school, and your driving habits for ways to cut back. There are many checklists and formulas to help you spot the places where heat, electricity, or gasoline are being wasted. In a practical text with clear diagrams the author explains the entire energy picture, where we went wrong in the past, what amount and kinds of energy we can count on in the future, and how we can learn to use this energy responsibly.

SAVE THAT ENERGY

ROBERT GARDNER

Illustrated

Julian Messner   New York

Published by Julian Messner, a Simon & Schuster
Division of Gulf & Western Corporation,
Simon & Schuster Building,
1230 Avenue of the Americas,
New York, New York 10020.
JULIAN MESSNER and colophon are trademarks of
Simon & Schuster, registered in the U.S. Patent
and Trademark Office.
The drawings in this book are by Henifer Mead.

Manufactured in the United States of America.
Design by Irving Perkins Associates

Library of Congress Cataloging in Publication Data

Gardner, Robert, 1929–
Save that energy.
Bibliography: p.
Includes index.
SUMMARY: Discusses the problem of diminishing
energy sources and ways of conserving energy and
eliminating its waste in the home, in school, and
in driving.
1. Energy conservation—Juvenile literature.
[1. Energy conservation] I. Title.
TJ163.35.G37   333.79′.16   80-28527
ISBN 0-671-34066-2

CONTENTS

PREFACE

DURING THE month of July, 1979, I spent three weeks at Providence College, Rhode Island, where I attended an institute on residential energy management sponsored by the United States Department of Energy.

One concern among the staff and teachers at the Providence institute was how to communicate information about energy and its conservation to a wide segment of the American people. It was there that I first began to think about this book. While class discussions of energy, courses on energy, and energy workshops conducted by the thirty teachers attending the institute might reach a few hundred students and adults, I realized that a book on energy, written for young people, could bring an understanding of energy issues and the methods of energy conservation to literally thousands of America's future energy consumers. This book is addressed to that task.

I hope that those of you who read this book will gain a better understanding of the energy issues that confront our nation and world and that you will discover some practical ways to conserve energy and reduce the amount of money your family spends to heat your home, run your appliances, and fuel your automobile.

SAVE
THAT
ENERGY

CHAPTER

1

ENERGY: A GLOBAL VIEW

WHAT WOULD happen if there were no coal, oil, or natural gas? How would your life change? To begin with, you would have to read this book in the daytime because there would be no electric lights at night. Most electric power plants produce electricity with steam turbines. The steam to drive the turbines is produced by heating water with burning coal, fuel oil, or natural gas. Without electricity or gas, your refrigerator would not work, nor would your stove, furnace, or air conditioner. There would be no street lights, no subway, no traffic lights, and no traffic. There would be no fuel to power automobiles, buses, trains, or airplanes. No fuel to heat homes, stores, schools, and factories.

We seldom realize how much we depend on electricity and fuels until a crisis arises. In the early evening of November 11, 1965, most of the northeastern United States was suddenly plunged into darkness. It was the largest power failure and blackout in history. People were trapped in elevators, subways, and office buildings. Vital machinery stopped; planes in flight had to land by moonlight because there were no runway lights. Even the lights in the control panels of many power stations went out. A large segment of the country was at a standstill.

Since 1973 oil shortages have sometimes caused long lines at gasoline stations. Some states have had to institute methods of rationing the fuel for several months.

Our complex, urban society is powered largely by the energy stored in oil (or its products—kerosene, gasoline, and fuel oil), natural gas, and coal. Discussions of the energy crisis and sources of energy are favorite topics for TV and radio programs, editorials, and articles in newspapers and magazines, but have you ever wondered, what is energy?

Energy is a very complicated subject, but, basically, it is energy that enables us to get jobs done—to lift heavy weights, to make things move, to light a lamp, to heat a house, or to make electricity. We obtain the energy to do such jobs from various sources: oil, coal, natural gas, waterfalls, even sunlight. Of course, we do many jobs with electricity, but electricity is a secondary source of energy because is is produced from other energy sources. We can boil water by burning natural gas, but we can also boil it on an electric stove using electricity generated by burning gas in a power plant. The burning gas would be a primary energy source; the electricity is a secondary source.

Energy is elusive. We can't see it. We can't smell or feel it. It's a subtle idea. Energy has many forms. Motion is one form of energy. You've heard of other kinds too. Heat, light, electricity, and atomic energy are a few. All of these forms of energy are closely related to the idea of work.

WORK

To you, work probably means doing something you don't like, such as washing dishes or doing homework, but to a scientist, work has a very different and very definite meaning. It involves two things: (1) a force and (2) the distance that force pushes or pulls something. Work is measured by multiplying the force by the distance through which the force moves:

work = force $\times$ distance, or $W = F \times D$

If you lift one-half pound to a height of one foot, you do half as much work as you do when you lift one pound through one foot. Similarly, you do twice as much work lifting one pound through two feet as you do lifting the same weight one foot.

Because work involves both force *and* distance, you do no work at all if you push against a brick wall, because the wall doesn't move.

If you exert a force of 10 pounds to push a box 10 feet across a floor, you do an amount of work equal to 100 foot-pounds

Work = 10 pounds $\times$ 10 feet = 100 foot-pounds.

POWER

If you walk up a flight of stairs, you do some work on yourself. If you run up the stairs, you do the same amount of work, but you do it faster. The rate of doing work is called *power*.

$$\text{Power} = \frac{\text{work}}{\text{time}}$$

Power can be measured in foot-pounds per second, or any units of force times distance divided by time. A common unit of power you probably know of is *horsepower* (h.p.). One horsepower is 550 foot-pounds of work per second, or 746 watts (0.746 kilowatt).

In 1783 James Watt, the Scottish inventor, first used horsepower to measure the rate of doing work. He simply defined one horsepower as 33,000 foot-pounds per minute (550 foot-pounds per second) because he found that a good strong horse could work at about this rate for short periods of time. (To see how much horsepower you can develop, try Experiment 1 at the end of this chapter.)

Table I shows the power developed by an average person doing different kinds of work. As you can see, a person can "work like a horse" for only short periods of time:

TABLE I

KIND OF WORK	PERIOD OF TIME	HORSEPOWER DEVELOPED
Running upstairs	10 sec.	0.95
Climbing a treadmill	30 sec.	0.65
Mountain climbing (steep)	1 hr.	0.20
Mountain climbing (normal)	all day	0.10

Of course, many machines are much more powerful than horses:

TABLE II

POWER SOURCE	POWER DEVELOPED
Horse	0.6 h.p.
Volkswagen (Beetle)	48 h.p.
Piper Cub airplane	65 h.p.
Lindbergh's "Spirit of St. Louis"	223 h.p.
Douglass DC-3 airplane	Two 2,400 h.p. engines
Boeing 747 airliner	Four 164,000 h.p. jet engines (at 375 mph)
S.S. *Oceanic* cruise ship	60,000 h.p.

ENERGY AGAIN

When work is done on something, that something acquires *energy*. Because that something has energy, it, in turn, is often able to do work on something else.

A pile driver consists of a large weight, or ram, which is lifted and then dropped. The falling weight is used to drive long poles, called *piles*, into the ground. When the ram is lifted, energy is stored in it; therefore, it can do work when it falls.

You do some work when you wind a watch or a clock. The coiled spring inside the timepiece acquires energy as you tighten it. The spring is then able to do work on the gears inside. The gears, in turn, do work on the hour, minute, and second hands, making them move.

A wound watch spring is an example of *potential energy*. Energy is stored in the spring, so the spring has the potential to do work. The same is true of the stretched rubber band in a slingshot, the compressed air in a BB gun, an elevated pile driver, the chemicals in a flashlight battery, even the chemicals in the food we eat.

If we push on an object through a distance, we do work on the object and it moves. The speed it acquires depends on how much work we do on it. The energy associated with motion is called *kinetic energy*. When water flows over a dam, gravity pulls the water downward, giving it more and more kinetic energy.

The kinetic energy (KE) in a moving object is equal to one-half its mass (m) multiplied by its speed (v) squared:

$$\text{Kinetic energy} = \tfrac{1}{2} \text{ mass} \times \text{speed} \times \text{speed or}$$
$$KE = \tfrac{1}{2} mv^2$$

A car moving at a speed of 50 mph has four times as much kinetic energy as an identical one traveling 25 mph. (Remember—to find the kinetic energy you have to square the speed.) The faster car can do four times as much work as the slower car if it collides with something. Similarly, stopping the faster car will require four times as much work as stopping the slower one. If the drivers of both cars lock their brakes, the friction of the road on the tires will do work on the cars and stop them. But the faster car will skid four times as far as the slower one. That's why speeding is so dangerous!

CONSERVATION OF ENERGY

To make a pendulum swing you do some work on it. You raise the bob a small height as you pull it to the side (you give it some potential energy). When you release it, it swings downward, gaining kinetic energy as gravity works on it. Beyond the midpoint of its swing, the bob loses speed but acquires potential energy again. The potential energy it gains at the end of its swing is very nearly equal to the potential energy that you gave it.

Of course, the pendulum eventually stops. The energy seems to disappear, but careful examination shows that the missing energy can be accounted for by a very slight increase in the temperature of the string and surrounding air. Some heat was created by friction on the string and collisions of air molecules with the moving bob. This increased kinetic energy of molecules is known as *thermal energy*, or, more commonly, *heat*.

Energy may be transferred or transformed from one kind to another, but experiments show that it is never lost. Energy is never created or destroyed—it is always conserved.*

Energy is changing constantly from one form to another. In a pendulum clock the potential energy in the raised weights is converted to the kinetic energy of the gears and the swinging bob. The chemical energy in the food you eat is transformed into kinetic energy when you run, or into gravitational potential energy when you pedal your bike up a hill. The potential energy of water is changed to kinetic energy as the water flows down a river. If the water falls over a dam, its increased kinetic energy may be used to turn a turbine connected to a generator and produce electrical energy.

* So you see you can't help conserving energy! When most people say, "Conserve energy," they mean use as little energy as possible to get jobs done.

In all these energy transfers, some heat is produced. Friction within your body, between your bike and the road, and between flowing water and the earth, the dam, and the turbine all produce some heat. It is for this reason that inventors who tried to produce perpetual-motion machines never succeeded. And they never will! We know now that such a machine is impossible. Whenever energy is transferred, a little of the energy rubs off as heat. If we convert potential energy to work, we never get as much work out as we put in; a little heat is always produced in the process.

Electric motors can lift weights, pump liquids, compress gases, and even provide the power to move automobiles. Because electricity can be used to do all this work, it, too, is a form of energy.

The round-and-round motion of a turntable, an electric mixer, or a power drill; the back-and-forth motion of a sander or an electric knife; the up-and-down motion of an elevator or a saber saw—all these motions are powered by the energy from electric motors.

When you turn on an electric mixer, a power drill, or an electric light, the electric energy comes from a power plant. In the power plant, giant generators consisting of huge wire coils turning in magnetic fields produce the electricity. The coils are connected to turbines that can be turned in a number of ways. Some are rotated by the conversion of the potential energy in water at the top of a dam to kinetic energy as it falls. The moving water turns the turbine blades. In other power plants the turbines are driven by jets of steam produced by heating water with burning oil or coal. An increasingly common

This high-pressure turbine (625 lbs./sq. in.) at the Connecticut Yankee Power Plant is turned by 7,340,961 pounds of steam per hour.

source of the heat used to generate steam comes from atomic reactors. The nuclear reaction does not create electricity; it simply produces heat that is used to make the steam needed to turn the turbines.

When coal, oil, and wood are burned, heat is produced. This heat was stored in the chemicals that make up these substances. This form of stored energy is *chemical energy*. It is the same kind of energy that is stored in the food we eat. Our bodies "burn" food slowly, producing the heat we need to stay warm, as well as the energy we need to carry on an active kinetic-energy-filled life.

But what is the source of the chemical energy stored in food, wood, and fossil fuels?

LIGHT ENERGY

You can tell that light is a form of energy because it can be changed to heat just as other kinds of energy can. But unlike other sources of energy, light comes to us free of charge from the sun.

Under the right conditions light can be converted to chemical energy as well as to heat. This means that light energy can be stored as a form of potential energy.

The green plants that cover much of the earth's surface (in the oceans as well as on land) are able to "capture" light and store it as chemical energy in the food they manufacture. This process is called *photosynthesis*. During this process the green plants use the energy in sunlight to convert carbon dioxide and water into food and oxygen. Since the food contains more energy than the carbon dioxide and water from which it was made, and since photosynthesis cannot take place in the dark, light must be the source of this extra energy.

The oxygen produced during photosynthesis is the same gas we breathe, the gas that most living things need to carry on respiration.

Since only green plants are capable of carrying on this magnificent process, the earth's entire source of food and oxygen comes from these plants. Without them, life would be impossible.

The chemical energy in the coal and oil we use to produce heat and electrical energy came from sunlight too. But the conversion of light energy to the chemical energy stored in these fuels took place eons ago. Coal and oil are the decomposed remains of plants and animals that lived on earth millions of years before humans appeared on this planet.

Of course, plants are still decomposing, but the conversion of dead plants into coal and oil takes a very long time. During the last hundred years we have used a large amount of the coal and oil that were formed over the last billion years.

MEASURING ENERGY

Just as distance can be measured in units such as inches, feet, yards, and miles, or in millimeters, centimeters, meters, and kilometers, so energy is measured in many different units. In this book we shall generally use units called Btu's (British thermal units). One Btu is equal to the amount of heat required to raise the temperature of one pound of water through one degree Fahrenheit.

Different people prefer to use different units. Nutritionists use the Calorie (the amount of heat required to raise the temperature of one kilogram of water one degree Celsius); most scientists use the joule (the amount of heat required to raise the temperature of one gram of water through 0.24 degrees Celsius); a few use the calorie (small c) which is the heat required to raise the temperature of one gram of water one degree Celsius; engineers use both Btu's and foot-pounds; power companies charge us for kilowatt-hours (3.6 million joules).

Table III will enable you to convert energy from one unit to another and to determine the energy stored in

large amounts of different energy sources. You may want
to refer to this table in this and later chapters.

TABLE III

TO CONVERT ENERGY IN UNITS OF	TO UNITS OF	MULTIPLY BY
kilowatt-hour (kwh)	Btu	3,413
Btu	kwh	0.000293
Calorie	Btu	3.96
Btu	Calorie	0.252
calorie	Btu	.00396
Btu	calorie	252
joule	Btu	0.000952
Btu	joule	1,050
quad	Btu	1,000,000,000,000,000 (10^{15})
1 ton of bituminous coal	Btu	26,200,000
1 barrel (bbl) of crude oil	Btu	5,600,000
1 gallon of gasoline	Btu	125,000
1 gallon of #2 fuel oil	Btu	138,800
1 cubic foot of natural gas	Btu	1,031
1 cord of wood (avg. density)	Btu	15,000,000

THE ENERGY PROBLEM

The energy problem is worldwide. The major industrial
nations of Western Europe, Japan, and the United States,
as well as many small developing nations, use more energy
than they produce from their own resources. To supple-

ment their energy sources, these nations import huge quantities of oil from the Middle East, Africa, and Latin America. The United States alone imported 8.5 million barrels per day in 1977—about half of the oil we consumed that year.

Because these nations spend so much money on imported oil, they often have an unfavorable balance of payments; that is, the amount of money spent on imported goods, particularly oil, exceeds the amount of money received from exports. This economic problem has arisen because the price of oil has skyrocketed in recent years. The OPEC (Organization of Petroleum Exporting Countries) nations (Nigeria, Iran, Kuwait, Qatar, Iraq, Saudi Arabia, Libya, Algeria, Ecuador, Gabon, United Arab Emirates, Venezuela, and Indonesia) who control the world's major oil resources have consistently agreed to keep the price of oil at inflated levels for nearly a decade. Because oil is the primary energy source in most industrial nations, the price of oil has been a major force driving the inflation that has swept the world.

The impact of OPEC pricing has been severely felt in the United States, where one-third of the world's energy is consumed by less than 6 percent of the world's population. A nation that developed in an age of cheap and abundant energy has been jolted by the sudden rise in energy costs. We have been slow to develop both the conservation measures required to reduce our use of energy and the alternate sources of energy needed to replace oil. Both approaches are essential if we are to reduce our demand for foreign oil.

For years, faith that geologists would be able to find

new sources of oil kept industrial nations from seeking and developing alternate sources of energy. Oil was cheap; it was abundant; it was easy to transport; it could be refined into a variety of products; and it required relatively little energy to extract and refine it. Now, a growing world population with more and more countries demanding more and more energy to fuel their developing industries and support their improving standards of living has made it clear that the energy resources of this globe are limited and cannot meet the demands of a bulging world forever.

From 1960 to 1973 the world's energy consumption grew from 130 quads (130,000,000,000,000,000 Btu's) to over 250 quads. The use of oil swelled from 33 percent to 40 percent of the total energy sources consumed during the same period. The combined share of oil and natural gas as an energy source increased from 48 percent to 66 percent, while coal declined from 47 percent to 28 percent. Despite coal's reduced share in the world's energy supply over this period, its annual consumption increased by nearly 1 percent.

During these same years the United States increased its use of energy from 47 quads to 74 quads. Our consumption of oil increased from 25 to 35 quads; the combined use of oil and natural gas rose from 34 quads (71 percent of total energy used) to 56 quads (76 percent of total). At the same time, the imported fraction of our oil and natural gas grew from 14 percent to 23 percent. Our own oil production increased 37 percent from 1960 to 1970 and then declined by 4 percent from 1970 to 1973 as our imports grew.

The New England states have been particularly affected by rising energy costs during the past several years. Some 80 percent of New England's energy needs are supplied by oil, compared with 47 percent for the nation as a whole. Furthermore, 80 percent of that oil is imported, so nearly two-thirds of New England's energy production depends on imported oil.

WORLD OIL AND COAL RESOURCES

The world is not going to run out of oil this year or during this decade, but the end of oil as a major energy source will probably occur within a century. It took millions of years for decaying plants and animals to form the vast quantities of oil, coal, and natural gas that existed beneath the surface of the earth a century ago. Within 200 years an energy-hungry world will use up the oil that required millions of years to form.

We cannot wait millions of years for our oil supply to be replenished, so we will have to seek alternative sources of energy. The obvious alternative is coal. Our coal deposits greatly exceed our oil supplies. There is probably enough coal to last another two to four hundred years. But coal is more difficult to extract and use than oil. Coal mining is a dangerous occupation, and burning coal seriously pollutes the atmosphere.

As you can see from the graphs of coal and oil resources, the world has coal resources in excess of 7,600 billion tons; its oil resources total only about 2,100 billion barrels. From Table III it can be determined that one ton of coal contains nearly five times as much energy as a

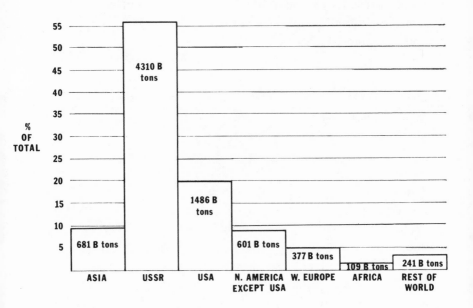

COAL RESOURCES

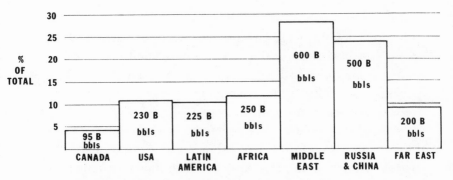

OIL RESOURCES

World's coal and oil resources

barrel of oil; consequently, the energy stored in the world's coal is more than seventeen times that stored in its oil resources.

Unfortunately, for every three units of energy that we obtain from coal, we must spend one unit of energy to extract and transport the material. The net energy ratio for oil is much better: 6.5 to 1.0 (we get 6.5 units of energy for every one unit invested).

How long will the world's oil supply last? We have already used 400 billion barrels of the total 2,100 billion barrels, and the rate at which we consume oil continues to increase. Table IV shows that the lifetime of the world's remaining oil depends on the rate at which our demand for oil increases in the future. We will assume a total supply of 2,100 billion barrels less the 400 billion barrels already used. Current world oil production is about 22 billion barrels per year.

TABLE IV

GROWTH RATE (% INCREASE IN USE PER YEAR)	REMAINING LIFETIME OF THE WORLD'S OIL AT THIS % GROWTH RATE (YEARS)
0	70
1	53
3	38
5	30
7	25
10	19

To illustrate why the lifetime shrinks so drastically with increasing growth rate, the 10 percent rate is shown year by year in Table V.

A growth rate of 10 percent would deplete our oil

supply in less than twenty years. Notice that it would take nearly fourteen years to use half the oil, but less than six years to use the second half. Clearly, it is very important that the growth rate of oil consumption be kept as low as possible, particularly if we continue to depend on oil as an energy source in the future.

TABLE V

Year	Oil remaining in earth (billions of bbls)	Oil used during the year (billions of bbls)	Increased use over previous year due to 10% growth rate (billions of bbls)
0	1,600	22.0	0
1	1,578	24.2	2.2
2	1,554	26.6	2.4
3	1,527	29.3	2.7
4	1,498	32.2	2.9
5	1,466	35.4	3.2
6	1,430	38.9	3.5
7	1,392	42.8	3.9
8	1,349	47.1	4.3
9	1,302	51.8	4.7
10	1,250	57.0	5.2
11	1,193	62.7	5.7
12	1,130	69.0	6.3
13	1,061	75.9	6.9
14	978	83.5	7.6
15	886	91.9	8.4
16	785	101.1	9.2
17	673	111.2	10.1
18	269	148.0	13.5
19	106	162.8	14.8
20	−73	179.1	16.3

ENERGY: A COMPLEX PROBLEM

There are alternatives to oil as an energy source. The world has over 7.5 trillion tons of coal, and we have used only 1 percent of it. Sunlight brings us abundant energy, but it is diffuse and difficult to concentrate for practical use. (We are still learning how to use solar energy.) Some scientists believe that hydrogen fusion (the process that takes place during a hydrogen bomb explosion) will provide a nearly inexhaustible supply of energy when we learn to control the process. This step will probably become possible early in the next century, but a controlled hydrogen fusion process that releases more energy than it consumes has yet to be demonstrated.

For the rest of this century and perhaps for several decades into the next century, our limited sources of oil must supply a significant fraction of our energy. This need will further complicate the economic, technical, and political problems associated with oil. Despite diminishing oil supplies, the demand for energy will increase to meet the needs of the world's increasing population. Industrial nations will strive to expand their productivity and increase their exports. Developing countries will seek more energy to fuel their growing industries and improve their standards of living.

Dwindling oil resources coupled with growing energy demands will enable OPEC to continue to increase the price of oil. Rising energy prices will cause production costs to soar; therefore, the cost of living will maintain its upward spiral, leading to a decline in the standard of living as people feel the effects of worldwide inflation.

Energy-deficient nations will continue to reel under the impact of an unfavorable balance of trade that causes money to flow to OPEC countries. (From 1970 to 1979 the money spent on imported oil by the United States increased from five billion to sixty-five billion dollars.) To meet their energy needs, many nations, particularly the United States, may turn to coal. Unfortunately, the production and transportation of coal are expensive; coal cannot be used as a fuel for automobiles, buses, trucks, airplanes, or modern trains; and its waste products seriously pollute the atmosphere. The polluting substances released during the combustion of coal can be removed from the waste gases; however, such procedures will increase the costs of production and contribute to the inflationary spiral.

Because both coal and oil release carbon dioxide, increasing use of these fuels will contribute to the current buildup of this gas in the earth's atmosphere. Carbon dioxide reflects back to earth heat that would otherwise escape into space. Many scientists believe the air's increased carbon dioxide content will create what is called a "greenhouse effect" that will raise the earth's average temperature. You might think that such a change in temperature would be welcomed because it would reduce homeowners' needs for heating fuels and thereby decrease the world's energy demands. Unfortunately, a small rise in the earth's temperature would melt the polar ice caps. This melted ice would raise the level of the oceans, leaving many coastal cities under water. Just move your finger along the coastal areas of a map of the world. Imagine how the world would change if New York, New Orleans,

Miami, San Francisco, London, Buenos Aires, Rio de Janeiro, Naples, Lisbon, Sydney, and other great seaports were beneath the sea.

Increased use of nuclear power as an energy source would reduce air pollution, but the disposal of radioactive wastes from atomic power plants is a problem that must be solved first. In addition, public concern about the safety of such plants following the incident in 1979 at Three Mile Island will impede the expansion of nuclear power.

ENERGY CONSERVATION

As you have seen, energy is a very complex problem, one that will require considerable thought, research, and compromise in the years ahead. As an energy consumer and potential homeowner, you must realize that the days of cheap energy have passed. It is in your home or apartment and the other dwellings of America that part of the energy crisis must be met and solved. Each of us, each member of every family, *can* reduce the energy we use in our homes and on the highways. By so doing, we will accomplish two things. First, we will reduce the total energy requirements of this country and our dependence on foreign oil. Second, we will keep our personal energy costs from soaring to a level that would require a significant decline in our standard of living.

In later chapters you will discover many ways to reduce energy consumption and, thereby, your family's energy costs. Conserving energy will not only save you and your family money, but it will also help your country reduce its dependence on foreign oil and improve its balance of payments.

EXPERIMENTS

1. WORKING LIKE A HORSE

Have someone measure the time it takes you to run up a flight of stairs. Then measure the vertical distance between the first floor and the second floor. If you know how much you weigh, you can figure out how much power you developed:

$$\text{power} = \frac{(\text{your weight}) \times (\text{height between floors})}{\text{time}}$$

How much horsepower can you develop?

2. AN ELECTRIC MOTOR

To build a simple electric motor, you will need a magnetic compass, a flashlight battery, and a long length of insulated wire. Wind the insulated wire around the compass so that it is parallel with the compass needle as shown in the drawing on page 49.

Leave several inches of wire at each end of the wire coil so that you can connect the coil to the battery. (You will have to sandpaper or strip the insulation from the ends of the wire to make good electrical contact between the battery and the wire.)

Tape one end of the wire to one pole of the battery. Briefly touch the other pole of the battery with the other end of the wire. What happens to the compass needle when electricity flows through the coil? By touching the wire to the battery and then removing it at just the right time you can get the compass needle to turn around and around. You have made a simple electric motor.

2

ENERGY SOURCES:
TODAY AND TOMORROW

Millions of years ago the earth's climate was very warm and humid. Plants flourished and so did many kinds of animals including now extinct dinosaurs. As these plants and animals died, they formed thick layers of decaying matter. The buildup of these layers produced pressures and high temperatures that squeezed out moisture and generated chemical changes leading to the formation of coal, oil, and natural gas. Because these fuels are the remains of ancient plants and animals, they are called *fossil fuels*.

FOSSIL FUELS

More than 90 percent of our nation's energy comes from fossil fuels. In 1976 some 47.3 percent of the energy used in this country came from oil. This oil was refined (separated by distillation) into gasoline, fuel oil, kerosene, petroleum gases such as propane and butane, and heavy tarlike liquids. In the same year coal supplied 18.5 percent of our energy needs, and 27.4 percent came from natural gas.

Nearly 60 percent of the oil we consume (25 percent of our total energy use) is used to transport people and materials in cars, trucks, buses, trains, and planes. Another 18 percent is used to heat homes and stores. Almost 15 percent is consumed by the industrial sector of the economy.

About 35 percent of our natural gas consumption is used to heat homes and businesses; 45 percent is consumed by industry. About 20 percent of the coal dug from United States soil is used in industry; a large fraction of this is required to manufacture steel. Some 13 percent of our coal is exported, and 4 percent is stored. The rest of our coal, 15 to 20 percent of our natural gas, and 10 percent of the oil we consume are used to generate electricity.

ELECTRICITY

Most American homes depend on the electricity produced in power plants where giant turbines turn electric generators. The generators consist of huge coils of wire that are rotated in magnetic fields to produce electricity.

About 10 percent of our electric power comes from turbines turned by steam generated from nuclear energy; 3 percent comes from hydro power (the kinetic energy in flowing water), and the rest from fossil fuels.

In this small hydroelectric power plant in Falls Village, Connecticut, the kinetic energy of water is converted to electricity.

Where will our electricity come from as our supply of fossil fuels dwindles?

There are a number of potential sources of energy that may drive power plants in the future, but none is well developed today.

The rhythmic motion of the ocean tides is one possibility, but this perpetual ebb and flow of power would be difficult to tap. In North America, the Bay of Fundy in Canada and Cook Inlet in Alaska are the only two sites where we could control the flow of water in large bays. To transmit power from these remote sites to centers of population would result in large losses of energy. High tides are rhythmical, but they do not match our peak energy demands because a high tide that occurs at 6 o'clock on Monday will arrive at about 6:50 on Tuesday. Contruction costs for tidal power plants would be huge, and, finally, such plants would inhibit the movement of fish and interfere with the ecology of the bays and estuaries behind these plants.

In Clarksboro, New Jersey, Vito Maglio's 60-foot-high windmill converts the wind's kinetic energy into 50,000 kilowatt-hours of electricity each year. Because the wind doesn't blow constantly, Mr. Maglio has to have storage batteries to provide electricity during calm air periods. These batteries can be recharged by the windmill when it is producing more energy than Mr. Maglio can use. In fact, this windmill produces about twice as much energy as the Maglio family needs. A recent invention, the synchromous inverter, is used to synchronize the frequency of the AC output of the windmill's generator with that of the utility line. In this way, Mr. Maglio is able to sell electricity to the power company.

Windmill enthusiasts believe that by the year 2000 windpower will be producing the energy equivalent of 850,000 barrels of oil per day (well under 5 percent of our probable energy demand by that time).

We are a trash-producing society. Americans discard five pounds of solid waste per person each day—nearly a ton per person per year. Why not use this annual 200 million tons of trash as a substitute for fossil fuels? It certainly makes more sense than burning it in an incinerator or covering it with gravel at a landfill site. In Denmark, plants that generate electricity from solid waste have been in operation for forty years. These power plants are usually built in cities to reduce the cost of transporting trash and to permit co-generation—the use of waste heat from condensing steam to heat nearby buildings.

Before you decide that solid wastes can replace fossil fuels as an energy source, consider the following problems:

- Solid waste must be collected and transported to the power plant.
- About 40 percent of the trash won't burn and must be separated before or after combustion.
- Although it is profitable to separate metallic waste for recycling, the separation process requires energy, and useless noncombustible materials must be discarded.
- Combustible plastics in our trash produce poisonous gases when they burn.
- The refuse must be processed to convert it to a form that can be handled and fed into steam-generating boilers.

Some scientists and engineers suggest that refuse should first be converted to solid, liquid, and gaseous fuels before being used in power plants. Two approaches being considered are pyrolysis and bioconversion.

Pyrolysis is the destructive distillation of waste materials. When substances such as paper and wood are heated in a closed oxygen-free container, they decompose into gases, liquids, and a charcoal-like solid. The gas produced is rich in methane, carbon dioxide, and hydrogen. It emerges from the heated container through a cooling tube, is collected, and stored. The thick, oily liquid—a mixture of organic chemicals rich in wood alcohol—condenses and separates from the gases as they pass through the cooling tube.

Researchers estimate a profit of one dollar per ton of refuse if the gases produced by pyrolysis are used to fuel the pyrolytic process. About a barrel of energy-rich liquid (10,500 Btu's per pound) and 160 pounds of charcoal can be collected and sold for each ton of refuse processed.

Bioconversion is simply the natural bacterial decomposition of organic matter that occurs during the decay of dead organisms and waste. If the process takes place in a closed container, gaseous products rich in methane are produced along with a sludge having a volume about 40 percent that of the original waste. Unfortunately, the methane is mixed with water vapor and carbon dioxide. These noncombustible gases have to be separated from the methane before it can serve as a fuel.

The sludge can be dried and sold as a fertilizer or animal feed because it is rich in nitrogen; however, the sludge is laced with trace amounts of copper, cadmium, and mercury that could be toxic to plants or animals.

Another form of biconversion, fermentation, is being used to produce gasohol. The sugar in a grain such as corn is changed to alcohol when it is mixed with yeast in an oxygen-free environment. When the alcohol content reaches 14 percent, the yeast cannot survive. The alcohol is then separated from the rest of the liquid mixture by distillation.

Gasohol is made by mixing alcohol with gasoline in a 1 to 9 ratio. Although gasohol is an effective fuel for cars, it will not be used to generate electricity. It may, however, ease the demand for gasoline and allow oil refineries to reduce their production.

The ocean waters off the coasts of Florida, Hawaii, and Puerto Rico all have the potential to produce electric power. The surface temperature of ocean water in the tropics and subtropics may be 80 degrees Fahrenheit, but several thousand feet below lies denser 40-degree water. This large temperature difference can be used to generate electricity. A low-boiling fluid such as ammonia (B.P. $= -28°F$) will boil in tubes lying in the ocean's warm surface. The vapor will turn turbines in a great floating power plant. The vapor exhaust from the turbines will enter a condenser where, under high pressure, cold water pumped up from the ocean's depths will condense the gas to a liquid so that the cycle may begin again. (The critical temperature of ammonia—the maximum temperature at which the fluid can exist as a liquid under pressure—is 271 degrees Fahrenheit, so ammonia can be condensed at 40 degrees, a temperature 68 degrees above its boiling point, provided the pressure is great enough.)

A small pilot plant to test this method is already operating off the coast of Hawaii. There is hope that at least some electricity will be produced in this manner during the next century, but it is difficult to predict the environmental effects of this system.

Another primary source of energy, one that costs nothing to mine or drill, is sunlight. In fact, solar energy is the source of other energy forms, for it is the sun warming the earth's atmosphere that creates the air currents we call winds. It is the sun warming the ocean's surface that creates the temperature differences needed for ocean thermal power. As you read earlier, it was the sun that provided the chemical energy stored in the world's fossil fuels.

Because sunlight is free, nonpolluting, and, in many places, plentiful, it may become at least a partial solution to our energy crisis. By using solar energy to heat buildings and the water we use for bathing and washing clothes and dishes, we can reduce our need for energy from coal, oil, and natural gas. Eventually we may be able to produce electricity from sunlight. Today the photovoltaic cells used to convert sunlight to electricity are so expensive and inefficient that electricity generated by sunlight is limited to satellites and remote areas of the world.

Nuclear power plants already provide about 10 percent of our electricity. Some people believe a much larger percentage of our electricity will come from *atomic energy* in the next century.

Aerial view of the Connecticut Yankee atomic power plant in Haddam.

When coal burns, atoms of carbon combine with atoms of oxygen to form molecules of carbon dioxide gas. Because heat is released during this process, we know that there is less chemical energy in carbon dioxide than there was in the carbon and oxygen before they combined. The loss of potential energy that appears as heat during the burning of coal might well be called *atomic energy* since it comes from the energy stored in atoms. However, the term *atomic energy*, as it is commonly used, refers to a much more potent source of energy, the energy released during the fission or fusion of atoms.

Atoms consist of two parts. There is a tiny, dense, positively charged core called the *nucleus* that contains protons and neutrons and that has practically all of the atom's mass. Protons and neutrons have the same mass, but neutrons have no charge. Protons each carry one unit of positive charge. The second part of the atom is made up of electrons moving around the nucleus at distances ten thousand times or more the diameter of the nucleus. While an electron has only one two-thousandth the mass of a proton, it carries one unit of negative charge. Because an atom has equal numbers of protons and electrons, its overall charge is zero.

All uranium atoms have ninety-two protons in their nuclei, but some have different numbers of neutrons than others. These different kinds of uranium atoms are called *isotopes*. (Many other elements share this property of having atomic nuclei with different numbers of neutrons.) Under the right conditions, uranium nuclei with 143 neutrons (uranium 235) will split to form lighter nuclei. During this process, called *fission,* several neutrons are released. These escaping neutrons can collide with other uranium nuclei with 143 neutrons, causing them to fission. If enough uranium nuclei are present, the neutrons released by one uranium nucleus can split two more uranium nuclei, the neutrons from these two can split four other nuclei, the neutrons from these four can split eight nuclei, and so on. Once this chain reaction starts, it takes but a fraction of a second before a huge number of uranium atoms undergo fission.

The smaller nuclei of such atoms as barium and krypton that are produced when a uranium atom splits have less total mass than the original uranium nucleus. The missing mass is transformed into energy. The amount of energy released is equal to the missing mass multiplied by the speed of light squared. This can be summarized by Einstein's now famous equation: $E = mc^2$. Because the speed of light is so large—186,000 miles (300,000 kilometers) per second—a small decrease in mass releases an enormous quantity of energy.

When an atomic bomb explodes, a vast number of uranium nuclei are fissioned, releasing huge amounts of energy. Such explosions are extremely destructive and hardly the solution to our energy crisis. However, it is possible to build atomic reactors in which the rate of fission can be controlled by surrounding small rods rich in uranium 235 with graphite to slow neutrons and by inserting boron steel rods to absorb neutrons. In this way energy can be released slowly enough to heat water to steam, as in conventional power plants.

A number of atomic power plants have already been built in various places throughout the world. They are generally found near large bodies of water because water is needed to cool the reactor and to provide a source of steam. These reactors do not produce carbon dioxide, nor do they pollute the air, but many people do not believe that nuclear fission is the solution to our energy crisis. Atomic power plants are very expensive to build; there is always the possibility of an accident that might release dangerous radioactive materials into the environment, and the problem of disposing of the radioactive wastes that accumulate in the plants during the fissioning process has yet to be solved. Furthermore, the wastes will have to be transported from the power plant to a burial area. An accident along the way might release dangerous radiation.

A hydrogen bomb explosion is an even more potent source of energy than uranium fission is. When a hydrogen bomb goes off, it is again a loss of mass that accounts for the enormous amount of energy released. However, the decrease in mass occurs during a process called *fusion*. During fusion, nuclei of hydrogen fuse (join) to form nuclei of helium. The helium nuclei have slightly less mass than the total mass of the hydrogen nuclei that combined to form them; the missing mass is transformed into energy.

Scientists and engineers are trying to find a way, using lasers or some other technique, to produce the high temperatures needed to set off a fusion reaction and at the same time control the rate of energy release. If they succeed, fusion might well solve the energy crisis for ages to come. There is certainly an abundance of the hydrogen isotopes used in fusion in the waters of our oceans. But the problems involved in igniting and controlling fusion are difficult and numerous. Controlled fusion power is probably at least twenty years away, and it will take at least one or two decades after that before it will have any major impact as an energy source for society.

Of course, we have a gigantic source of fusion power only 93,000,000 miles away—our sun. There, more than 650,000,000 tons of hydrogen is converted to helium every second. The resulting helium is several tons less massive than the hydrogen from which it comes, and the missing mass appears as the energy emitted by the sun. Fusion, then, is the source of solar energy.

SYNTHETIC FUELS

The decomposition of organic materials or trash by micro-organisms or heat is one of several ways scientists plan to synthesize fuels. A number of pilot plants are now investigating the possibilities of (1) converting solid coal to combustible liquids and gases that can be transported and burned far more easily than coal, (2) extracting oil from shale rock or tar sands, (3) developing a more efficient way to pump the heavy oil that is now left in the earth to the surface where it can be refined.

Fluid fuels may be obtained from coal by pyrolysis, indirect liquefaction, or direct liquefaction. When coal is heated in an oxygen-free environment, three fuels are produced: a high energy gas called pipeline gas (a mixture of gaseous hydrocarbons); liquid syncrude; and char, a charcoal-like residue. Syncrude, like crude oil, is a mixture of liquid hydrocarbons that can be refined into gasoline, kerosene, and fuel oil. Because of the paucity of hydrogen in coal, only 1¼ barrels of syncrude can be obtained from a ton of coal.

Indirect liquefaction converts coal to hydrogen and carbon monoxide by passing steam over hot coal. By the use of high pressure and catalysts, these combustible gases are converted to a liquid.

Direct liquefaction uses a catalyst, high pressure (270 times atmospheric pressure), high temperatures, and the addition of hydrogen to produce 2½ barrels of liquid per ton of coal.

One hundred billion barrels of shale oil could probably be squeezed from limestone rocks (marl) in Colorado, Utah, and Wyoming. A ton of marl must be mined, pulverized, and squeezed at high temperatures to obtain two-thirds of a barrel of oil. The environmental effects of strip mining marl and spewing vast amounts of dust into a region already plagued by dust storms, however, makes the process abhorrent to many people.

A pilot plant in Alberta, Canada, is producing 45,000 barrels of oil by heating and squeezing tar sands. Tar sands there are believed to hold another 100 billion barrels, a quantity equal to all of the oil produced in the United States since 1859 when "Colonel" E. L. Drake drilled America's first oil well in Titusville, Pennsylvania.

At least 30 billion barrels of heavy oil lie beneath United States soil. An estimated trillion barrels can be found along the Orinoco River in Venezuela. As light oil resources are depleted, the extraction of heavy oil will become economically competitive.

Although all of these synfuels may become increasingly important energy sources in the future, none of them is currently feasible. Some even have a negative energy balance; that is, more energy is used in producing the fuel than can be obtained from it. In addition, the use of each of these fuels has the same bad features as our present fossil fuels.

EXPERIMENTS

1. PYROLYSIS OF PAPER

Cut a sheet of newspaper or wooden coffee stirring sticks into small pieces and pack the pieces into a test tube. Using a one-holed and a two-holed rubber stopper, glass tubing, a beaker or tumbler, another test tube, a bottle, rubber tubing, clamps, and a bucket of water, set up the experiment shown in the drawing.

Heat the paper-packed test tube with an alcohol burner. (*Be careful with the fire and be sure to wear safety glasses when you heat glass.*) Gases emerging from the

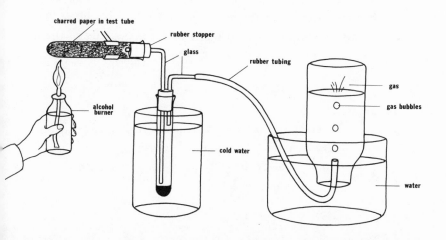

hot test tube will condense in the test tube immersed in ice water if they boil at temperatures greater than 32 degrees Fahrenheit. More volatile gases will collect in the inverted water-filled bottle.

Once gas bubbles begin to collect in the bottle, you might like to test the flammability of the gas. Remove the rubber tubing from between the bottle and the tube where gases are condensing and use a match to see if you can ignite the gas emerging from the cold test tube.

Reconnect the tubing and collect some more gas. Move the burner along the paper-filled test tube to decompose as much of the paper as possible. When the bubbling rate in the bottle diminishes, remove the tube from the bottle and the bucket. Place a cover over the mouth of the bottle and remove it from the water. Will the gas in the bottle burn?

Continue to heat the test tube until all of the paper appears to be black.

The oily liquid in the cold test tube is a mixture of many liquids—water, wood alcohol, and several hydrocarbons. How can you tell that more than one liquid has condensed?

When the paper-filled test tube is cool, remove the black solid that remains. What does it look like? Hold one of the black flakes or chips with forceps and heat it in a flame. Does it burn or does it glow?

2. Generating Electricity

To see that electricity is produced when you change the magnetic field through a coil of wire, you will need a galvanometer—a device that can detect electricity. To

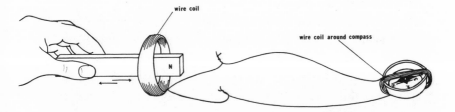

make one, wrap about 20 feet of enameled copper wire around a magnetic compass. Be sure to wind all the wire in the same direction. Leave about a foot at each end to make electrical connections. Use sandpaper or a knife to remove the insulating enamel from an inch or two of wire at each end of the coil that you have made around the compass.

You can see that the galvanometer responds to electricity by touching the two bare ends of the coil to opposite ends of a 1½-volt battery.

Make a second coil of wire by wrapping another 20 feet of wire around one of your hands. Use a little tape to hold the wires together. Again, sand the two ends of the coiled wire.

Connect the ends of the two coils of wire with two long pieces of wire as shown in the drawing. Be sure to sand the ends of the wires if you use enameled wires to connect the two coils.

Be certain that the compass needle of the galvanometer is parallel to the wire in the coil around it before you bring a magnet near the other coil. (The second coil should be far enough from the galvanometer so that the magnet alone does not affect the compass needle.) What happens to the galvanometer when you move the magnet in and out of the coil? What happens if you turn or move the coil near the end of the magnet?

As you can see, any change in the magnetic field within the wire coil induces an electric current. By turning very large coils in strong magnetic fields, a great amount of electric energy can be produced. You know that the energy to turn such coils generally comes from steam turbines.

3

SAVE
THAT ELECTRICITY

MOST AMERICAN families buy electrical energy from a utility company. How does a utility company know how much to charge you? How can you reduce the amount of electricity you use?

Find the place where the electric power line enters your house or apartment. It will be connected to a meter like the one in the next drawing. Electricity flowing through the meter causes the disc inside to turn. The disc is connected through a gear train to the dials that measure the electrical energy entering your house. The dial on the right records kilowatt-hours, the next dial indicates tens of kilowatt-hours, the next hundreds of kilowatt-hours, and so forth.

The turning disc is similar to a car's speedometer. Its rate of turning indicates how fast electricity flows into

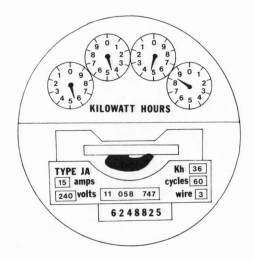

your home. The dials are like a car's odometer. Just as you can read the change in numbers on an odometer to find how far a car has traveled in a day, month, or year, so a representative from your utility company can read dials on the electric meter to determine how much electricity has flowed into your house since the last reading.

The meter dials in the drawing indicate 5,448 kilowatt-hours. If, after a month, the dials read 7,448, the utility company will charge your family for 2,000 kilowatt-hours. Assuming a flat rate of 5¢ per kilowatt-hour, your bill will be $100 ($0.05 × 2,000).

WHAT'S A KILOWATT-HOUR?

You can think of electricity as tiny bundles of electric charge that move along the fine wires in light bulbs, toasters, and hair dryers, or the wire coils of electric motors in vacuum cleaners, refrigerators, washing machines, and blenders, or along the electric beam that illuminates the screen of your TV set.

Electric current is the rate at which these bundles of charge flow. If a lot of charge moves through your electric meter each second, the current is large. If few charges move past, the current is small.

Current can be measured with an ammeter in units called amperes. One ampere (1 A) is equal to a charge flow of one coulomb per second. A coulomb is the name we give to a bundle of charge. Actually, the bundle contains 6.25×10^{18} (6,250,000,000,000,000,000) electrons. Each charge carries some energy that can be measured with a voltmeter. A reading of one volt (1 V) on a voltmeter means that each coulomb of charge carries one joule of energy that will be converted to heat, light, motion, or some other form of energy as it flows along the circuit:

$$1\,V = 1\,\frac{J}{C}$$

Most household circuits operate at 120 volts; therefore, each coulomb of charge will deliver 120 joules of energy. (Electric stoves and dryers are usually on 240-volt circuits.) Suppose the current through a light bulb filament in an ordinary household circuit is 0.5 A (½ coulomb per second); the energy produced in the bulb will be 60 joules per second. Since the ammeter measures charge per second (current) and the voltmeter measures the energy per charge, the product of the two readings will give you the energy per second. In this case:

$$0.5\,A \times 120\,V = 0.5\,\frac{C}{s} \times 120\,\frac{J}{C} = 60\,\frac{J}{s}$$

It's much like finding distance traveled from speed and time. If you drive at 50 mph for two hours, you will travel 100 miles:

$$50 \, \frac{mi}{h} \times 2 \, h = 100 \, mi$$

The energy produced per second, you'll remember, is power and can be measured in watts. A bulb carrying a current of ½ ampere in a 120-volt circuit receives 60 joules per second; it will be rated and sold as a 60-watt bulb. A bulb that carries a current of one ampere is a 120-watt bulb: ($1 \, A \times 120 \, V = 120 \, W$)

If a 120-watt bulb burns for one hour (3,600 seconds), the energy produced will be 432,000 joules.

$$120 \, \frac{J}{s} \times 3,600 \, s = 432,000 \, J$$

Energy is equal to power multiplied by time. If your family is using lots of lights and appliances simultaneously, the total current through your meter might be 100 A. In one hour the energy produced in all these 120 volt circuits would be:

$$120 \, V \times 100 \, A \times 3,600 \, s = 43,200,000 \, J$$

To avoid such large numbers, utility companies measure power in kilowatts (kw) (1,000 watts = 1 kilowatt), and energy in kilowatt-hours (kwh). One kilowatt-hour is equal to 3,600,000 joules:

$$1 \, kwh = 1,000 \, w \times 3,600 \, s = 1,000 \, \frac{J}{s} \times 3,600 \, s = 3,600,000 \, J$$

The energy, in kilowatt-hours, produced in a 100-watt bulb that is on for one hour would be:

$$0.1 \, \text{kwh} \times 1 \, \text{h} = 0.1 \, \text{kwh}$$

A home that draws a current of 100 A on 120-V circuits for one hour would be charged for 12 kwh since:

$$120 \, \text{V} \times 100 \, \text{A} \times 1 \, \text{h} = 12{,}000 \, \text{w} \times 1 \, \text{h} =$$
$$12 \, \text{kw} \times 1 \, \text{h} = 12 \, \text{kwh}$$

A charge for 12 kwh is certainly an easier number to use, print, and read than 43,200,000 joules.

HOUSEHOLD APPLIANCES

Table I contains a list of electrical appliances commonly found in American homes. The first column of numbers contains the usual wattage rating for each appliance (a number that is generally printed on the appliance). The second column of numbers gives the *average* time each appliance is operated during a one-year period. The third column of numbers contains the product of the numbers in the first two columns; that is, the energy required to operate each appliance for one year. The last column lists the average cost to operate each appliance for a year assuming a cost of 5¢ per kilowatt-hour.

Do you think the average figures given are right for your home?

SAVE THAT ENERGY

TABLE I

APPLIANCE	AVERAGE WATTAGE	AVERAGE NUMBER OF HOURS USED PER YEAR	KILOWATT-HOURS PER YEAR	COST TO RUN PER YEAR ($)
Kitchen				
Blender	390	40	15.6	0.78
Broiler	1,400	70	98	4.90
Coffee maker	900	120	108	5.40
Dishwasher	1,200	300	360	18.00
Range	12,200	100	1,220	61.00
Microwave oven	1,450	130	189	9.45
Toaster	1,200	35	42	2.10
Waffle iron	1,100	20	22	1.10
Refrigerator (12 cu. ft.)	240	3,000	720	36.00
Frostless refrigerator (12 cu. ft.)	320	3,800	1,220	60.80
Freezer (15 cu. ft.)	340	3,500	1,190	59.50
Frostless freezer (15 cu. ft.)	440	4,000	1,760	88.00
Laundry				
Clothes dryer	4,800	200	960	48.00
Iron	1,000	140	140	7.00
Washing machine	500	200	100	5.00
Water heater	2,500	1,600	4,000	200.00
Quick-recovery water heater	4,500	1,000	4,500	225.00
Comfort				
Air conditioner	900	1,000	900	45.00
Electric blanket	180	830	149	7.45
Dehumidifier	250	1,500	375	18.75
Fan (attic)	370	800	296	14.80
Fan (window)	200	850	170	8.50
Humidifier	180	900	162	8.10

TABLE I *continued*

APPLIANCE	AVERAGE WATTAGE	AVERAGE NUMBER OF HOURS USED PER YEAR	KILOWATT- HOURS PER YEAR	COST TO RUN PER YEAR ($)
Health and Beauty				
Hair dryer	750	50	37.5	1.90
Shaver	14	80	1.12	0.06
Sun lamp	280	60	16.8	0.84
Toothbrush	7	60	0.42	0.02
Entertainment				
Radio	70	1,200	84	4.20
TV (B/W, tube)	160	2,200	350	17.60
TV (B/W, solid-state)	55	2,200	121	6.05
TV (color, tube)	300	2,200	660	33.00
TV (color, solid-state)	200	2,200	440	27.00
Housewares				
Clock	2	8,760	17.5	0.88
Vacuum cleaner	630	75	47	2.36
Sewing machine	75	140	10.5	5.25
Lighting				
Light bulbs (on, in home)	660	1,500	1,000	50.00

WAYS TO SAVE ELECTRICITY

The price of electricity will increase because, for the next few years, it will be produced by burning coal, oil, and natural gas, and you know what's happened to the price of these fossil fuels. Some utility companies offer special rates for electrical energy during off-peak hours. It's difficult and inefficient to close down a power plant; yet there's far less demand for electricity at night so some power companies offer very low rates at such times. You can reduce your electrical bill significantly by using electrical appliances during special rate times.

Here are some other ways to reduce your use of electricity in different areas of your home.

IN THE KITCHEN

- Your electric range is an energy gobbler. You can reduce its energy consumption in the following ways:

  a./ Cover pans when you cook food or boil water.

  b./ Use pans that cover the heating element to ensure that heat flows into the pans, not into the air.

  c./ Turn off heating elements and ovens several minutes before food is thoroughly cooked. The heat stored in the elements and ovens will then be used for cooking.

  d./ Use a timer when food is cooking in an oven. Heat is lost when an oven is opened, and the time required for cooking increases.

  e./ Cook as many foods as possible in the same hot oven. It takes nearly as much energy to cook one casserole as to cook six or seven foods simultaneously.

  f./ Cook on heating elements whenever possible. These units require less energy than the oven or broiler.

- An automatic dishwasher should be used only when filled with dishes. After the rinse cycle, let the dishes air dry.

- Use cold water or a pan of warm water when scraping dishes. Don't use a running stream of hot water.
- Keep your refrigerator set at 40 degrees Fahrenheit. Colder temperatures are not needed and waste energy.
- Be sure the gasket around your refrigerator door is tight. It should hold a dollar bill firmly.
- Before buying a new refrigerator-freezer, compare the energy needed to operate various models. Frostless models use about 60 percent more energy than those that require manual defrosting. The refrigerator should have an energy-saving switch and should be well insulated.
- A freezer should be operated at zero degrees Fahrenheit. It should be kept full to justify its cost. Buy food in quantity at sales, and store it in the freezer.
- If you're buying a freezer, consider a chest type that is well insulated. It will lose less cold air when opened than an upright model will.

IN THE LAUNDRY

- Wash clothes in cold or warm water, if possible. Rinse them in cold water.
- Operate washers and dryers only for full loads (but don't overload).
- Soak very soiled clothing before washing.
- Dry large amounts of clothes in consecutive loads so that the dryer's mass has to be heated only once.
- Don't run a dryer longer than necessary.
- Clean the lint screen after drying each load.

- Clean the outside vent once a month.
- Dry clothes outside on a line whenever possible.
- Wear wash-and-wear clothing as much as possible to reduce the energy required for ironing and dry cleaning.
- Insulate hot water pipes.
- Add insulation to old water heaters. Kits are available for this purpose.
- Turn off the water heater when your family is going to be away for more than a day.
- Turn the thermostat setting on your hot water tank to a lower temperature. This will reduce heat losses from tank to surroundings.

AROUND THE HOUSE

- Take quick showers rather than baths to reduce hot water use.
- Turn off lights, TV, radios, and other appliances that are not being used.
- Remove all but one bulb in multibulb fixtures.
- Use one bright bulb rather than several less powerful ones where bright light is essential.
- Keep light bulbs and fixtures clean; dirt absorbs light.
- Use four-watt night lights.
- Turn off outside lights when they are not needed.
- Use fluorescent rather than incandescent lighting wherever possible. Fluorescents are much more efficient.
- Buy solid-state TV sets. They use one-third less energy than comparable tube sets. Buy as small a screen as possible for the site. Large screens require more energy.

- If your heat is electric, keep thermostats set at 65 and turn them back to 50 or 55 at night or when a room is not being used.
- If a room is seldom used in winter, turn off its thermostat provided there's no danger of pipes freezing.

EXPERIMENTS

1. PANS—COVERED OR OPEN?

When you boil water on a stove, should you cover the pan? You can answer this question, in terms of energy, by finding out how long it takes to bring a cup of water to boil when you heat it in a covered pan, and then in the same pan without a cover. In both cases you should start with pan, water, and burner at room temperature. What do you find?

Suppose the cover were very heavy. Would this make a difference?

There is another factor to consider when you heat water on a stove. How much water remains in the uncovered pan when it starts to boil? How much remains in the covered pan when it reaches the boiling point?

As you can see, you could start with less than a cup of water in the covered pan and have as much water as you have in the uncovered pan when both reach boiling.

2. LIGHTING—FLUORESCENT OR INCANDESCENT?

You probably have both fluorescent and incandescent lights in your home. Is one more efficient than the other?

A bulb's power rating is written on it. The wattage

tells you how much electrical energy is developed in the bulb each second. A 100-watt bulb, for instance, generates 100 joules (0.095 Btu) of energy each second; however, not all of the electrical energy is converted to light. Carefully touch an incandescent bulb and a fluorescent bulb that have the same wattage ratings and have been burning for a few minutes. Which bulb do you think converts more electrical energy to heat? Which bulb do you think produces more light per watt?

To test your ideas, check the packages in which both bulbs are sold. The package provides information about the light emitted by the bulb, as measured in lumens, as well as its wattage rating.

A lumen is the amount of visible light that falls on an area of one square foot at a distance of one foot from a one-candle light source. The reading on a light meter under such conditions would be one footcandle.

I have a 60-watt incandescent light on my desk. The package in which it came indicates that the bulb produces an average of 850 lumens. Its efficiency in lumens per watt would be $850/60 = 14.2$. Now you can check the efficiency of various incandescent and fluorescent bulbs. How does the efficiency of incandescent bulbs change as the wattage increases? How does the efficiency of incandescent bulbs compare with that of fluorescent bulbs of the same wattage?

To compare the efficiency of fluorescent and incandescent lights experimentally, you can use a light meter or the exposure meter on a camera to measure the light emitted by the two types of bulbs.

Stand several yards (meters) from an incandescent

bulb in an otherwise dark room. Measure the light intensity (illuminance) or exposure setting. Replace the incandescent bulb with a fluorescent bulb of the same wattage, and again measure the light intensity at exactly the same distance. (For a camera keep the diaphragm adjustment or f-stop fixed and change the shutter speed to find the proper exposure. If the shutter speed has to be doubled, the intensity of the light must have doubled as well.) Which bulb gives more visible light per watt?

For Science Experts: Cut a one-inch square from a sheet of cardboard. Place the square hole on the incandescent bulb so that the light from one square inch of the bulb is emitted in your direction. (Be sure that only the light from the one-square-inch hole reaches you.) Measure the light intensity at different distances from the bulb. Repeat the experiment using the fluorescent bulb.

What happens to the light intensity as you move farther from the light source? At the same distance, how does the light intensity from the fluorescent bulb compare with that from the incandescent bulb? *But,* how does the *total area* of the fluorescent bulb compare with the total area of the incandescent bulb? Which bulb emits more light? How many times as much?

3. Hot Water—An Energy Drain

Hot water accounts for about 20 percent of the average homeowner's energy bill. You can estimate the cost of hot water in your home quite easily. The next chart shows you how many gallons of hot water, on the average, are required for each of the jobs listed. Estimate the number of times each job is done in your home every week. Multiply this estimate by the number of gallons required for the job, and you will have the weekly hot water requirement for that job. Add up the hot water needed for all the jobs, and you will have your weekly hot water requirements.

Job	Number of gallons to do the job	Number of times the job is done per week	Gallons of hot water per week
Load of laundry	20	_____	_____
Bath	25	_____	_____
Shower	10	_____	_____
Dishwasher	15	_____	_____
Washing dishes in sink	2	_____	_____
	Total hot water volume per week =		_____

An average household uses 350 gallons of hot water per week. Is your family above average, average, or are you already conserving hot water?

How much hot water would your family use in a year? (Assume you use the same amount each week for all 52 weeks.)

It takes about 0.14 kilowatt-hour to heat a gallon of water to 120 degrees Fahrenheit. How much energy does it take to meet the hot water needs of your family for one year?

To find out what it costs your family for hot water, multiply the number of kilowatt-hours required by the cost of a kilowatt-hour. For instance, if it requires 2,500 kwh to supply hot water, and the cost of electricity is 5¢ per kilowatt-hour, the energy costs for hot water will be $2,500 \times 0.05 = \$125$.

The above calculation assumes that the water loses no heat once it has been warmed to 120 degrees. Of course, this is not true, especially if the tank is poorly insulated. To reduce heat losses from your tank, here are some suggestions:

- Adding insulation to the tank can reduce costs by as much as 50 percent.
- Lowering the thermostat on your hot water tank to 105 degrees will reduce heat losses by about 30 percent because the rate of heat flow is proportional to the temperature difference between the hot water and its surroundings.
- Repair leaky faucets.
- Insulate hot water pipes.
- Turn off your hot water tank when your family goes away for a day or more.
- If your utility company charges lower rates during off-peak hours, turn off your hot water tank except during those hours. Try to adjust your use of hot water to take advantage of these rates.
- Install a flow reducer in your shower head.

4. SHOWER OR BATH

How much hot water do you use to take a bath? To take a shower?

Next time you draw your bath water, measure the

depth of the water in the tub. The volume of water in the tub (in cubic inches) can be found by multiplying the area of the water by its depth. For instance, if the area of the water surface is 1,000 square inches and the depth is 6 inches, the volume of water in the tub is 6,000 cubic inches. A gallon is 231 cubic inches. How many gallons of water did you draw for your bath?

Before you take your next shower, close the drain so you can measure the volume of water you use. Try to shower reasonably quickly. After you step out of the tub, measure the depth of the water.

How does the amount of hot water used for a shower compare with the amount used for a bath? How does the cost of a quick shower compare with the cost of a bath?

One-fifth of a kilowatt-hour is needed to heat a gallon of water. If you heat with gas, about a cubic foot is required; an ounce of fuel oil will do the same job. How much does it cost to take a bath? To take a shower?

5. ENERGY CONSERVATION

From past utility bills, determine how much electricity, gas, and other energy sources your family uses each month. Then lower the thermostats, set thermostats back at night, reduce the use of lights, appliances, and hot water. Compare your family's use of energy over the next few months with utility bills from the previous year. How much have you reduced your family's use of energy? How much money have you saved by these conservation measures? (Compare what your family pays with what they would have had to pay if they hadn't reduced their use of energy.) Why is a comparison of this year's energy bills with last year's a poor way to estimate energy savings?

6. SETTING BACK THERMOSTATS

One way to reduce energy use during the winter is to turn back the thermostat(s) in your home to 65 degrees in the daytime and 50 degrees at night or when your family is away. If you haven't done this already, you can test the impact of this conservation measure in your home quite easily.

First, determine your weekly or monthly consumpion of energy by reading your oil, gas, or electric meters at weekly or monthly intervals. If you use gas or electricity for heat and other jobs that require energy, compare winter bills with summer bills. The difference is probably your heating cost. Set your thermostat back for the heating season and compare last year's energy use with this year's. To obtain a more accurate comparison, divide the energy used for heating by the number of degree-days for each year. (See chapter 4.) How could you reduce energy use in the summer if you have air conditioning?

7. ENERGY USE AND APPLIANCES

To find the energy required to operate any electrical appliance in your home, you need to know two things: its wattage rating and the length of time the appliance operates. You can find the wattage rating printed on the appliance. If the appliance is large (a stove or refrigerator), the manual that came with the appliance will supply the information, or you can call a local appliance dealer and obtain the information you need. In some cases you will find a plate that gives the current and voltage supplied to the appliance. From that information how can you determine the wattage rating?

Next, you will have to estimate the time each appliance is used during the course of a year. By multiplying the kilowatt rating and the length of time (in hours) the appliance operates, you can determine the energy needed to run the appliance for a year. Suppose the light on your desk has a 100-watt bulb and you use it an average of two hours a day for 365 days; that's 730 hours. One hundred watts is 0.10 kilowatt, so the energy required to operate your desk lamp for one year is 73 kilowatt-hours (0.10 kw × 730 h). If you pay 5¢ per kilowatt-hour, it costs $3.65 ($0.05 × 73) to operate that light for one year.

If you do your homework at your desk, your family would probably agree that $3.65 is a good investment, provided you turn off all other lights in the room.

Make a survey of all the appliances in your home. How much energy and money are needed to operate each one for one year? See how closely your figures agree with those in Table I. Can you suggest ways to help your family reduce the use of appliances? How much money can be saved if they follow your suggestions?

AN ELECTRICAL ENERGY QUIZ

Consult Table I on pp. 56–57 to answer some of these questions.

1. How much energy could the average family save each year if they watched color TV (tube) half as much as they do now? How much money could they save?

2. How much energy could the average family save if they used a clothesline rather than a dryer? How much money would they save?

3. By lowering hot water thermostats to 120 degrees and adding insulation to the outside of the tank, most families can save about $70 per year on their electrical bills. How much energy will they save? By what percent will they cut their hot water costs?

4. You can reduce the operating cost of a dishwasher by one-third if you open the door after the final rinse and let the dishes air dry. How much energy would you save each year by doing this? How much money?

5. How much energy would a family save in one year by *not* buying a frostless refrigerator to replace the manual one they own?

6. By taking showers instead of baths the average family can save about 2,000 gallons of hot water a year. It takes 400 kilowatt-hours to heat this much water. How much money would the average family save if they switched to showers? (Savings in energy and money can be made even greater by using shower heads that restrict the water flow and by reducing time spent in the shower.)

7. Fluorescent lights give five times as much light

as incandescent bulbs of the same wattage. How much energy could be saved by the average family if they substituted fluorescent lights for half their incandescent lights? How much money would they save?

8. An attic fan requires about 300 kilowatt-hours of energy per year to operate. If the average family could use such a fan instead of an air conditioner to cool their home, how much energy could they save? How much money would they save?

9. Do you think most families would consider it worthwhile to use wind-up mechanical clocks in place of their electric clocks?

10. Do you think you could convince someone that he could save money by replacing his electric razor with a blade razor? Do you think he would save energy? Remember that most men shave with hot water.

11. What is the power rating of a 120-volt toaster that draws a current of 10 amperes?

12. The toaster operates 10 minutes (⅙ hour) each morning. How much energy will it deliver in kilowatt-hours? In joules?

13. If electricity costs 5¢ per kilowatt-hour, how much will it cost to operate the toaster each morning?

4

SAVE
THAT HEAT

SEVENTY-SEVEN million billion Btu's of energy were generated in the United States in 1978. That's 77 quads. One-fifth of that energy, 15 quads, was used in American homes. Half of those 15 quads were used for space heating.

If you could turn on your furnace, warm your home to 70 degrees, and then turn off the heat for the rest of the winter, very little energy would be required. Unfortunately, it's not that simple. Heat continually flows out of buildings through ceilings, walls, windows, doors, floors, and foundations. Furnaces or electrical heating units must replace the heat that is lost to the cold air outside.

WHAT IS HEAT AND HOW DO WE MEASURE IT?

Early scientists thought of heat as an invisible fluid (they called it caloric) that flowed from warm bodies to colder ones. Fuels were somehow able to store caloric in a latent form that was released when the fuels burned. Today we believe that heat is the kinetic energy of the tiny molecules that make up matter. Fuels contain more chemical energy than the products formed when they burn. During burning, chemical potential energy is converted to the molecular kinetic energy we call heat.

Many people confuse heat and temperature. They are *not* the same. Temperature measures the *average* kinetic energy of the molecules of a substance. Heat is the *total* kinetic energy of all the molecules that make up a sample of material; therefore, the amount of heat in a substance depends on how much of the substance there is. The temperature doesn't. Two pounds of water (about a quart) will supply twice as much heat as one pound for each degree it cools.

To understand how heat is measured and how it differs from temperature, you can try the same experiments I did. I placed a pound (a pint) of water in an insulated container and used an electric immersion heater (the kind used to make a cup of hot water for tea or coffee) to warm the water. I plugged the heater into a wall outlet for one minute. After pulling the plug, I used the heater to stir the water while the water's temperature rose from 60 degrees to 75. I repeated the experiment using different

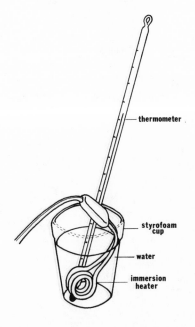

amounts of water and heat. I assumed that a heater plugged in for two minutes would deliver twice as much heat as it did when plugged in for one minute, and half as much heat if plugged in for only thirty seconds.

Table I is a summary of my data. One unit of heat equals the heat produced by the heater in one minute.

In the first set of experiments, one unit of heat was delivered to each of three different amounts of water. The results reveal that doubling the amount of water halves its temperature change; halving the amount of water doubles its temperature change.

In the second set of experiments, you see that when the heat supplied to a fixed amount of water is doubled, the temperature change of the water is also doubled.

SAVE THAT ENERGY

TABLE I

Heat supplied to water (units)	Weight of water used (lb)	Initial temperature of water (°F)	Final temperature of water (°F)	Change in temperature (°F)
1.0	1.0	60	75	15
1.0	2.0	60	67.5	7.5
1.0	0.5	60	90	30
0.5	1.0	60	67.5	7.5
1.0	1.0	60	75	15
2.0	1.0	60	90	30
1.0	1.0	60	75	15
2.0	2.0	60	75	15
3.0	3.0	60	75	15

The last three trials indicate that if the amount of heat and the volume of water are both doubled or tripled, the temperature change will be the same.

All of these experiments suggest that heat depends on *both* the amount of water and the change in the water's temperature. If we double the amount of heat to a fixed amount of water, we see that the temperature change is doubled; if we double the amount of heat, we can keep the temperature change constant by doubling the amount of water.

Examine Table II. It was prepared using the data in Table I. It shows that the *product* of the water's *weight and temperature change* doubles when the heat doubles.

Both weight *and* temperature change are needed to measure heat.

TABLE II

Heat Supplied (units)	Weight of Water (lb)	Temperature Change of Water (°f)	Product of Weight and Temperature Change (lb × °f)
1.0	1.0	15	15
1.0	2.0	7.5	15
0.5	1.0	7.5	7.5
0.5	0.5	15	7.5
2.0	1.0	30	30
2.0	2.0	15	30
3.0	1.0	45	45
3.0	3.0	15	45

We can use our knowledge that heat is related to both the weight and temperature change of water to define a unit of heat. Engineers define the Btu as the amount of heat required to raise the temperature of one pound of water through one degree Fahrenheit. In my experiment, the unit of heat was equal to 15 Btu's because in one minute the heater raised the temperature of one pound of water through 15 degrees.

Other units of heat are similarly defined. A calorie is the amount of heat needed to raise the temperature of one gram of water through one degree Celsius. A joule is the amount of heat required to raise the temperature of one gram of water through 0.24 degree Celsius.

Few substances can absorb as much heat per degree temperature change as water. When I repeated my experiment using one pound of cooking oil in place of water, the same unit of heat produced a temperature change of 30 degrees Fahrenheit. This means only half as much heat (0.5 Btu) is required to raise the temperature of one pound of cooking oil through one degree. The heat required to raise a pound of a substance through one degree is called the specific heat of that substance. Table III lists the specific heats of some common materials.

TABLE III

SUBSTANCE	SPECIFIC HEAT (BTU/LB/°F)	SUBSTANCE	SPECIFIC HEAT (BTU/LB/°F)
Water	1.0	Ice	0.487
Steel	0.12	Glass	0.18
Copper	0.092	Gypsum	0.26
Aluminum	0.214	Sand	0.191
Concrete	0.22	Glass wool	0.157
White pine	0.67	Air	0.24

Note that air, the fluid that must be heated to warm your home, has a specific heat of 0.24 Btu/lb/°F. Because a cubic foot of air weighs only 0.075 pound, it takes only 0.018 Btu to raise the temperature of one cubic foot of air through one degree Fahrenheit.

$$0.24 \frac{\text{Btu}}{\text{lb}/°\text{F}} \times 0.075 \frac{\text{lb}}{\text{cu. ft.}} = 0.018 \frac{\text{Btu}}{\text{cu. ft.} /°\text{F}}$$

HOW DOES HEAT ESCAPE?

The molecules of air in your house have a large average kinetic energy because the air is warm. These fast-moving molecules bump into the molecules on the surface of the walls in your home and transfer some kinetic energy to them. These molecules, in turn, transfer kinetic energy to molecules deeper in the wall. Eventually, kinetic energy is transferred by the molecules on the outer surface of the walls to slower moving molecules in the cold air outside. We call this type of heat transfer *conduction,* and we say that heat "flows" from warm objects to colder ones.

Infiltration of cold air through cracks around windows, doors, and sills forces warm air out other cracks. This type of heat transfer is called *convection.* There are also convection currents within a building as warm, light air rises over cooler, denser air; however, convection heat losses through ceilings, walls, and floors are generally minimal because the small air spaces within these structures break up convection currents.

Heat, like light, is transferred from the sun to earth by radiation. Because space is a virtual vacuum, the sun's heat could not reach us by conduction or convection. Standing near a large window on a cold day, you've sensed radiational effects. Heat radiates from your body to the cold window. You've also experienced radiant heat entering your body from a fireplace or when sunlight falls on your skin or clothing. If the ceilings, walls, and floors of a home are warm, relatively little heat will escape through radiation.

The two major causes of heat loss from a building are conduction and infiltration. Conduction losses can be reduced significantly by insulating walls, ceilings, and floors. Sealing the cracks that allow cold air to enter a house will eliminate drafts that result from infiltration.

Conductive heat losses are determined by four factors: time, surface area, the temperature difference between air inside and outside the building, and the insulating quality of the materials that surround the living space.

The time factor is obvious. If 20,000 Btu's escape from your house in one hour, 40,000 will escape in two hours if conditions don't change.

Heat, to escape, must pass through the surface of a building. If one house has twice as much surface area as another, twice as much heat will escape in the same time if all other factors are the same.

It should not surprise you to learn that heat is conducted from a building more rapidly as the outside temperature drops. The furnace goes on more often on a cold winter day than on a cool autumn afternoon. As you can see for yourself in Experiment 2 at the end of this chapter, heat flow from a warm body is doubled when the temperature difference between the warm body and its colder surroundings also doubles. This means that heat loss is proportional to the difference in temperature between a body and its surroundings, just as it is proportional to time and surface area.

Finally (and here's the factor we can control best), heat flow can be reduced by surrounding a warm body with insulation—materials that are poor conductors of heat.

To determine the heat conductivity of a material, we

measure how much heat flows through one square foot of its surface each hour for each degree of temperature difference between the enclosed warm body and its colder surroundings. Suppose a warm air space surrounded by 10 square feet of ¾-inch wood sheathing loses 100 Btu's of heat per hour when the temperature of the enclosed air is kept 10 degrees warmer than the air outside the wood. The conductivity of the wood sheathing is 100 Btu's per hour per 10 square feet per 10 degrees Fahrenheit, or 1 Btu / h / sq. ft. / °F. If fiberboard is substituted for sheathing, the heat losses drop to 50 Btu's and so the conductivity, or *U value*, is 0.5 Btu / h / sq. ft. / °F.

Homeowners want insulating materials that do *not* conduct heat well; consequently, insulating materials are rated according to their ability to resist heat flow. A material's ability to resist heat flow is known as its *R value*. The R value is just the inverse of a material's U value, or conductivity. The U value of ¾-inch fiberboard is:

$$0.5 \text{ Btu} / \text{h} / \text{sq. ft.} / °F;$$

its R value is $\dfrac{2 \text{ sq. ft.} - °F - h}{\text{Btu}}$

It's just the conductivity, or U value, turned upside down.

A material with a large R value is a good insulator. Fibrous glass wool blankets 3½ inches thick are commonly used to insulate walls. They have an R value of 11. These blankets would have a U value of $\frac{1}{11}$ (0.091) and would conduct heat $\frac{1}{11}$ as fast as ¾-inch wood sheathing.

Table IV lists the R values of some common building materials. As you can see, doubling the thickness of a material generally doubles its R value. Note that still air along walls, ceilings, and floors is an insulator.

TABLE IV

Material	Thickness (inches)	R value sq. ft. – H – °F btu
INSULATION:		
Blankets, Batts—Mineral wool	1	3.1
" " " "	3½	11
" " " "	6	19
" " " "	7	22
" " " "	12	38
LOOSE FILL INSULATION:		
Glass fiber	1	2.2
Rock wool	1	2.7
Cellulose fiber (paper)	1	3.7
Vermiculite	1	2.2
Perlite	1	2.7
Expanded polyurethane	1	5.9
Expanded polystyrene	1	4.7
" " (molded beads)	1	3.8
Polyisocyanurate sheathing (Aluminum foil on both sides)	1	8.0
BUILDING MATERIALS:		
Wood sheathing	¾	1.0
Plywood	½	0.63
Bevel-lapped siding	½	0.81
Asbestos board	¼	0.13
Gypsum board (Sheetrock)	⅜	0.32
Plywood panel	¼	0.31
Building paper	—	0.06
Vapor barrier (plastic)	—	0
Wood shingles	—	0.87
Asphalt shingles	—	0.44
Linoleum	—	0.08
Carpet with fiber pad	—	2.1
Hardwood floor	—	0.71

TABLE IV *continued*

MATERIAL	THICKNESS (INCHES)	R VALUE SQ. FT. – H – °F / BTU
WINDOWS AND DOORS:		
Single glazed window	—	1.0
Double glazed window	—	2.0
Exterior door (wood)	—	2.0
MASONRY:		
Concrete block	8	1.1
Concrete block (lightweight)	8	2.0
Brick, common	4	0.80
Concrete, poured	8	0.64
AIR FILM AND SPACES:		
Air space, bounded by building materials	¾ or more	0.9
Air space, bounded by aluminum foil	¾ or more	2.17
Air film on exterior surface	—	0.17
Air film on interior surface	—	0.68

CALCULATING CONDUCTIVE HEAT LOSSES

We know that heat losses due to conduction are proportional to time, surface area, conductivity, and the temperature difference between inside and outside air. All this can be summarized by the following equation:

Heat loss = Conductivity × surface area × time × temperature difference

or in symbols:

$$H = U \times A \times t \times [T(\text{inside}) - T(\text{outside})]$$

Since the insulating qualities of materials are usually given in R values rather than U values, we can rewrite this equation as:

$$H = \frac{1}{R} \times A \times t \times [T(\text{inside}) - T(\text{outside})]$$

This formula enables you to calculate conductive heat losses from a building. Suppose the exterior walls of a house have an R value of 14 and a surface area of 1,000 sq. ft. If the inside temperature is 65 degrees and the outside temperature is 35 degrees, the heat loss *through these walls* in one hour will be:

$$H = \frac{1}{14} \times 1000 \times 1 \times (65° - 35°) = 2{,}143 \text{ Btu.}$$

Similar calculations can be made for the floors and ceilings. But the temperature outside the house is continually changing. You could measure the outside temperature every hour, repeat the calculation for each hour of the heating season, and then add up all the heat losses to find the heat lost for the entire season. This would be a very tedious process, so heating engineers have developed an easier method. They came up with the idea of degree-days. As you will see, a degree-day (DD) is similar to man-days of work. A person working one day is said to do one man-day of work.

Generally a home doesn't have to be heated if the outside temperature is above 65 degrees Fahrenheit. When the average outdoor temperature drops below 65, degree-days start piling up. To figure out the number of degree-days, we determine the average temperature for a day. This temperature is subtracted from 65 and multiplied by one day.

If the average temperature for one day is 45 degrees, the heating requirements for that day are said to be 20 degree-days (20° × 1 day). If the average temperature during the next day is 35° (30 degrees below 65°F), the heating requirements for that day are 30 degree-days. The total for the two days would be 50 degree-days.

When the average temperature is 25 degrees, you will use twice as much fuel as you would in 45-degree weather.

There are many more degree-days during the winter than in the spring or fall. No degree-days are recorded in the summer when the average daily temperature stays above 65.

Table V gives the normal degree-days for each month of the heating season (September through May) in Chicago, Illinois. Table VI gives the total number of degree-days for a number of cities in the United States.

TABLE V

DEGREE-DAYS DURING HEATING SEASON IN CHICAGO, ILLINOIS

SEPT.	OCT.	NOV.	DEC.	JAN.	FEB.	MARCH	APRIL	MAY
81	326	753	1,113	1,209	1,044	890	480	211

TABLE VI

NORMAL ANNUAL DEGREE-DAYS

City	Degree-days	City	Degree-days
Atlanta, Ga.	2,961	Honolulu, Hawaii	0
Boston, Mass.	5,634	Los Angeles, Calif.	2,061
Buffalo, N.Y.	7,062	Miami, Fla.	214
Chicago, Ill.	6,155	Milwaukee, Wis.	7,635
Cleveland, Ohio	6,351	Nashville, Tenn.	3,578
Dallas, Tex.	2,363	New York, N.Y.	4,871
Detroit, Mich.	6,232	Pittsburgh, Pa.	5,987
Duluth, Minn.	10,000	Portland, Ore.	4,635
Fairbanks, Alaska	14,279	St. Louis, Mo.	4,900
Hartford, Conn.	6,235	Washington, D.C.	4,224
Helena, Mont.	8,129	Wichita, Kans.	4,620

You can find the number of degree-days per heating season in your locality by reading the weather column in a local newspaper or by calling a fuel supply company. The degree-day map can give you an approximate value for degree-days any place in the United States. If you multiply the number of degree-days by 24 hours, you will have the total *degree-hours* for your heating season. This value is the sum of all the products of time and temperature differences for the heating season. Using $24 \times$ DD, you can calculate the heat losses for the entire season:

$$H = \frac{1}{R} \times A \times 24 \times DD$$

Using our earlier example, a house with 1,000 square

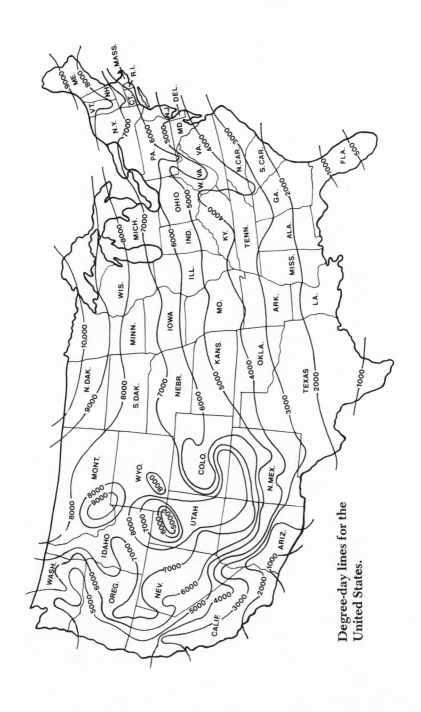

Degree-day lines for the United States.

feet of walls in a 5,000-DD city would lose about 8,600,000 Btu's:

$$H = \frac{1}{14} \times 1,000 \times 24 \times 5,000 = 8,571,428 \text{ Btu's.}$$

As you can see, nearly 8,600,000 Btu's of fuel will have to be used to replace the heat lost through these walls.

You can make similar calculations for heat losses through not only walls, but ceilings, floors, doors, and windows as well for your own home or school. Using Table VII, you can then make a reasonable estimate of how much your fuel bill will be for a heating season. Of course, prices in different areas at different times will vary, but you can substitute your values for the ones in this table and make calculations accordingly.

INFILTRATION HEAT LOSSES

Convection heat losses occur when cold air replaces warm by flowing through cracks around windows and doors, electrical outlets, foundations, or poorly sealed floors and ceilings. The exchange between inside and outside air is usually greater than is needed to supply fresh air. Of course, the rate of air turnover increases with the wind velocity.

Many stores or buildings where doors are constantly being opened have vestibules to reduce infiltration of cold air.

TABLE VII

HEAT SOURCE	COST PER UNIT ($/UNIT)	FUEL COST ($/M BTU)	INITIAL COST OF SOURCE	APPROXI-MATE EFFI-CIENCY (%)	COST FOR ACTUAL ENERGY DELIVERED ($/M BTU)	CONTROL OF HEAT SOURCE
Electricity	0.05/kwh	14.70	low	100	14.70	excellent
Oil	1.25/gal	8.90	moderate	70	12.70	very good
Coal	100/ton	3.80	moderate	60	6.30	fair
Wood	100/cord	6.66	moderate	50	13.30	fair
Natural gas	0.40/100 cu. ft	4.00	moderate	70	5.70	very good
Solar	0	0	high	low	low	poor

Heat losses due to infiltration are more difficult to calculate than those caused by conduction, but good estimates are possible if you know the number of air turnovers per hour, the volume of air in the building, and the temperature difference between inside and outside air.

Consider a house that has one air turnover per hour, 1,200 square feet of floor space, and 8-foot-high ceilings. On a day when the outside air temperature is 35 degrees, the volume of air that must be warmed from 35 to 65 degrees every hour is 9,600 cubic feet (1,200 sq. ft. $\times$ 8 ft.). Since 0.018 Btu's are required to warm one cubic foot of air through one degree Fahrenheit, the total heat needed to warm this infiltrating air each hour is:

$$H = 9,600 \text{ cu. ft.} \times 0.018 \frac{\text{Btu}}{\text{cu. ft.} - \text{h} - \text{°F}} \times 30° \text{ F} \times 1\text{h} = 5,184 \text{ Btu's}$$

In general, seasonal infiltration heat losses are given by this equation:

$$H = V \times 0.018 \times N \times 24 \times DD$$

(V is the volume of air per turnover, and N is the number of turnovers per hour.)

If the house referred to above is in an area where there are 5,000 degree-days per season, the infiltration heat losses for a season would be:

$$H = 9,600 \times 0.018 \times 1 \times 24 \times 5,000 = 20,736,000 \text{ Btu's}$$

REDUCING INFILTRATION

Converting a drafty house to a "tight" house can reduce N in the equation given and make an amazing difference in fuel costs. The number of air turnovers per hour is an estimate that can be made through careful observation. Table VIII on the next page shows the observations that must be made according to the Retrotech method to make a reasonable estimate of N.

To estimate N (the number of air turnovers per hour) multiply checks in first column by 1, second column by 2, third column by 3. Add the numbers and divide by 4. (You can, of course, give ratings of 1.5, 2.5, or even 0.5.)

The most economical way to reduce heat losses in many homes is to reduce infiltration by caulking doors, windows, sills, and cracks in walls or foundation, and by weather-stripping all doors and windows. Caulking and weather-stripping materials are relatively inexpensive, and the installation requires little skill. Even for a home with average fitting windows and doors, an investment in caulking and weather-stripping will pay for itself in less than one season.

Explanations of how to caulk and weather-strip can be found in many books. Two inexpensive but instructive booklets are:

Morrison, James W., Editor. *The Complete Energy-Saving Home Improvement Guide.* New York: Arco Publishing Company, Inc., 1978.

TABLE VIII

PUT CHECK ($\sqrt{}$) IN APPROPRIATE BLANK (____).

PART OF THE BUILDING	1 ONE AIR CHANGE PER HOUR IF ALL BLANKS CHECKED	2 TWO AIR CHANGES PER HOUR IF ALL BLANKS CHECKED	3 THREE AIR CHANGES PER HOUR IF ALL BLANKS CHECKED
Basement or	___Tight, no cracks. Sills, windows, entrances caulked. Windows seal well.	___Some cracks, loose windows and entrances.	___Major cracks, poorly sealed entrances and windows.
Crawl Space	___Plywood floor, no trapdoor. Water sewer and electrical openings into house well sealed.	___Tongue-and-groove board floor. Reasonably tight seal on trapdoor and around utility pipes.	___Board floor. Loose fit on utility pipes entering house.
Windows	___Storm windows. Tight-fitting windows, weather-stripped.	___No storm windows. Tight-fitting windows, weather-stripped.	___No storm windows. Loose fitting windows. No weather-stripping.
Doors	___Storm doors. Tight-fitting doors, weather-stripped.	___Loose storm doors. Poorly fitting doors. No weather-stripping on doors.	___No storm doors. Poorly fitting doors. No weather-stripping.
Walls	___Joints between doors and windows and wall caulked. Building paper under siding.	___Caulking loose. Needs painting.	___Cracks around windows and doors. No building paper under siding.

In the Bank or up the Chimney. Washington, D.C.: U.S. Government Printing Office, 1977.

Here are some other ways to reduce infiltration losses. These will cost nothing.

- Keep all doors closed, including those into the garage, attic, and basement. Do *not* close attic vents above ceiling insulation. These openings are needed to provide the air currents that carry away the moisture that diffuses into the attic.
- Close your fireplace damper when the fireplace is not in use so that warm air cannot escape up the chimney.
- Use your fireplace on cool *not* cold days. If possible, close off other rooms while a fireplace fire burns, and turn down the thermostat.
- Use exhaust fans only when necessary.
- Keep all windows closed when the heating or air-conditioning system is on.
- Go in and out exterior doors as infrequently as possible.

REDUCING CONDUCTIVE HEAT LOSSES

To reduce the conduction of heat from your home, *insulate!* If you're building a new house, insulate to obtain the R value recommended on the R-value map of the United States, on the following page.

If your home is not insulated, start with the attic. Because warm air rises, the temperature difference be-

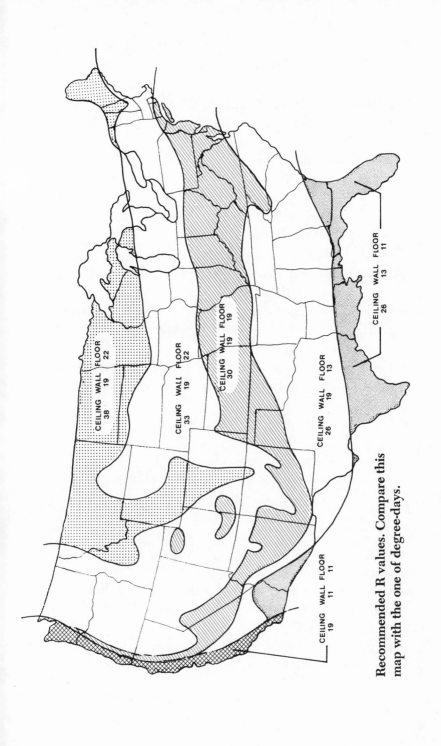

CEILING WALL FLOOR
38 19 22

CEILING WALL FLOOR
33 19 22

CEILING WALL FLOOR
30 19 19

CEILING WALL FLOOR
26 19 13

CEILING WALL FLOOR
26 13 11

CEILING WALL FLOOR
19 11 11

Recommended R values. Compare this map with the one of degree-days.

tween inside and outside air is greatest between ceiling and attic; consequently, conductive heat losses are greatest there. Six inches of insulation will increase the R value from about 1.7 to 20.7 and will reduce conductive heat losses to the attic by a factor of 12.

The addition of insulation to exterior walls, floors, and basement walls will also reduce heat losses significantly. To insulate the walls of an older home, insulation must be blown in. This is a job for a professional, but you can insulate ceilings, floors, and basement walls yourself.

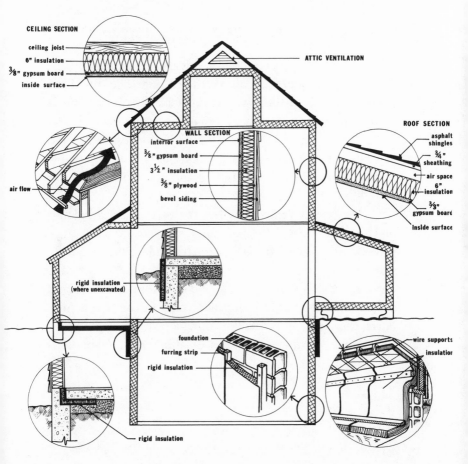

CEILING SECTION
ceiling joist
6" insulation
3/8" gypsum board
inside surface

ATTIC VENTILATION

WALL SECTION
interior surface
3/8" gypsum board
3½" insulation
3/8" plywood
bevel siding

air flow

ROOF SECTION
asphalt shingles
3/4" sheathing
air space
6" insulation
3/8" gypsum board
inside surface

rigid insulation
(where unexcavated)

foundation
furring strip
rigid insulation

wire supports
insulation

rigid insulation

For details about installing insulation, talk to your local lumber dealer or read the books just mentioned. The drawing indicates where insulation should be located in a house.

If insulation is not installed properly, it may create moisture problems. In winter the water vapor content in a cubic foot of warm inside air is usually greater than in the cold air outside; consequently, moisture tends to diffuse from inside to outside. If moisture condenses on the cold insulation, water fills the tiny air spaces that retard heat flow, and the insulation loses its effectiveness. To prevent this from happening, you should install a moisture barrier along with insulation.

In an attic, polyethylene strips can be stapled to the bottom of the joints just above the ceiling. The insulation is then placed on these vapor barriers. If this is difficult to do, interior ceilings can be made vapor-resistant by using two coats of oil-based paint. Vapor-resistant paint can also be used on exterior walls where it is impossible to install plastic barriers unless the sheetrock walls are torn off.

In all cases, remember that the vapor barrier belongs on the warm side of the insulation. It is also important to provide adequate ventilation on the cold side of the insulation to carry away any moisture that penetrates the barrier. Attic vents should *not* be closed, and you should be careful *not* to cover the soffit vents in the eaves. Similarly, there should be air flow outside exterior wall insulation. Don't fill the undersides of shingles or clapboards with paint.

The vapor barrier in both walls and ceilings is on the warm side. This building has 12 inches of insulation above the ceiling (R 38) and 6 inches in the wall (R 19). Note also the loose insulation around the window.

The builder was able to put 6 inches of insulation in the walls because he framed with 2 x 6-in. lumber instead of the usual 2 x 4-in. lumber.

TURN BACK THAT THERMOSTAT

You know that heat losses depend on the temperature difference between inside and outside air. Reduce that difference in temperature and you'll reduce heat losses, use less energy, and spend less money. Whether your home is well insulated or not, you can reduce heat losses by turning your thermostat to a lower setting. If the outside temperature is 45 degrees and you set your thermostat at 65, you will use only two-thirds as much heat as you would if the thermostat were set for 75. A setting of 55 would require only one-third the energy needed to maintain 75 degrees. A temperature of 55 degrees is uncomfortable for most people unless they are active, but it is certainly adequate at night when you are under layers of insulating blankets. Use an electric blanket if you want to; it's much less expensive than keeping an entire house at 65 degrees.

When no one is at home, it's foolish to keep the house at 65, and if a room is unoccupied there may be no reason to heat it at all.

Some people think that more energy is required to warm air back to 65 degrees than is saved when the thermostat is turned back to 50 or 55. Not so! Whatever the temperature inside, your heating system simply replaces heat that is lost from your home, and the heat lost depends on the temperature difference between inside and outside air. It's true that the furnace will run longer than usual to bring the temperature from 50 to 65 degrees, but that's more than compensated for by the time no heat was

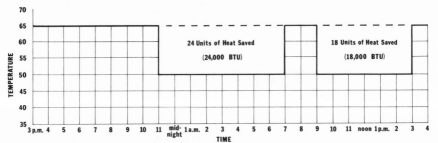

Setting thermostats back saves heat.

used as the house cooled from 65 to 50 and the even longer time that only enough fuel was used to replace heat losses from a 50-degree as opposed to a 65-degree house.

The graph should make it clear that setting back the thermostat does save energy. Each square represents a unit of heat lost from your house in one hour when the outside temperature is 35 degrees Fahrenheit. Let's say the unit is 1,000 Btu's for a well insulated house. If you keep the house at 35 degrees, it will cost nothing to heat it, but you'll be very uncomfortable. To keep the inside temperature at 40 degrees for one hour will require one unit of heat or 1,000 Btu's to replace the heat lost to the outside; a 45-degree inside temperature will require two units of heat, or 2,000 Btu's per hour; 50 degrees will require three units or 3,000 Btu's per hour, and so on. As the graph shows, turning your thermostat from 65 to 50 for six hours during the day when everyone is away will save 18 units of heat. Setting it back for eight hours at night will save 24 units of heat. The total saving of 42 units is nearly 30 percent of the energy that would be required to keep the house at 65 degrees for all 24 hours.

Translated into the cost of oil for a heating season,

it could mean a reduction of 30 percent in fuel bills if the average winter temperature is 35 degrees Fahrenheit. As you can see, even greater savings could be achieved in more moderate climates.

EXPERIMENTS

1. Heat Loss and Surface Area

Experience has probably convinced you that the rate at which heat flow from a body increases with surface area. If you wake up feeling cold while a blizzard rages outside, you'll feel warmer when you curl up. On the other hand (or season), you'll feel cooler on a hot, humid summer night if you stretch your body out across the bed.

Establishing the relationship between heat loss and surface area is not a simple matter because there are other factors that influence heat flow, such as insulating materials, the temperature difference between the warm body and its surroundings, and the time the heat flows. As with any experiment, you must be careful to test only one factor at a time; otherwise, you will not know which factor is producing the effect being observed, in this case, heat flow. Here is an experiment that will enable you to determine how heat flow is related to surface area.

Make two pieces of ice that have the *same* volume but different surface areas by pouring 30 to 50 milliliters (ml) (1 to 1½ ounces) of water into differently shaped containers. You might use a cubic vessel to make one ice cube and the top from a plastic container to make a pan-cake-shaped ice "cake" that has a large surface area. Or you could use two cylindrical containers with different diameters.

Because the ice cakes are made with the same volume of water, they weigh the same, and it will take the same amount of heat to melt each of them. To melt one gram of ice requires 80 calories, 334 joules, or 0.31 Btu's. To melt 30 ml (one ounce) of ice requires 2,400 calories, 10,000 joules, or 9.4 Btu's. (A milliliter of water weighs one gram, so you know the weight of the water you used to make the ice.)

Since you know the weights of the ice cakes, you will know how much heat flows into the ice to make it melt. Using the time required to melt the ice, you can determine how much heat flowed into the ice each second on the average. Because the ice cakes have different areas, you can compare the heat flow for equal times (one second) with the areas of the ice cakes.

To prevent the temperature difference between the cold ice and the warmer water from changing (which could affect the heat flow into the two pieces of ice), use a lot of water. The temperature of the ice will remain at 32 degrees Fahrenheit (0° C)° throughout the melting process. The temperature of the water will not change significantly, and the temperature of the ice is constant while it melts, so you can be sure that the temperature difference between water and ice will be constant during the time heat flows into the ice.

° If you don't believe this, put a lot of crushed ice into a container and measure the temperature. Then measure the temperature of a smaller volume of crushed ice. Be sure the thermometer bulb is well covered in both cases.

Once the ice cakes are thoroughly frozen, fill a bucket or large container with water. After measuring their dimensions add the ice cakes to the water. Stir the water constantly so the ice surfaces will be in contact with the warmer water and not their own melt water. How long did it take for each cake to melt? How much heat flowed into one of the ice cakes in one second? How much heat flowed into the other cake in one second? What was the surface area of each piece of ice?

Compare the following two ratios:

$$\frac{\text{Heat flow into ice cake } \#1 \text{ in one second}}{\text{Heat flow into ice cake } \#2 \text{ in one second}} \text{ and}$$

$$\frac{\text{Area of cake } \#1}{\text{Area of cake } \#2}$$

As you can see, heat flow is proportional to the surface area if the time and temperature differences between the ice and warm water are constant. This is why more heat per day will flow from a house with a large surface area than from one with the same volume that has less surface area.

2. RELATION OF HEAT FLOW TO THE TEMPERATURE DIFFERENCE BETWEEN TWO OBJECTS

Experience has taught you that the rate at which heat flows from your body depends on the temperature of the air around you. If you step outside on a cold day, you soon begin to shiver unless you are wearing warm clothing. Shivering is your body's way of asking your muscles to go to work and produce more heat to keep your body temperature from decreasing. To help you find out how heat flow is related to the temperature difference between two objects, here are two experiments.

A. *Heat Flow from Hot to Cold and Hot to Warm*

Add 100 ml (or ½ cup) of hot water to each of two styrofoam, plastic, or unwaxed paper cups. Measure the water temperature in each cup; they should be nearly the same. Leave one cup in a warm room and place the other in a cool place such as your refrigerator (or outdoors if it's cold). Record the temperature in each cup every minute. At some point in the experiment, record the temperature of the air around each cup.

In which environment does heat flow more rapidly?

To see how the cooling rates compare, plot a graph of temperature versus time for each cup. Plot temperature on the vertical axis and time on the horizontal axis. Both sets of results can be placed on the same graph. You might use different colors, or ink and pencil, to plot the data for the two cooling curves.

As you can see from your graphs, water loses heat faster in cold air. You might have noticed also that the cooling rate decreases as the temperature of the warm water becomes more like the temperature of the cooler surrounding air. This indicates that the *rate* of heat loss decreases as the temperature difference between two bodies drops.

For Science Experts: Use your graphs to determine approximately how fast heat is being lost by the warm water at any point on the graph. (You can measure heat in calories or Btu's. One milliliter of water weighs one gram, and one cup of water is about one-half pound.)

Plot a graph of the rate of heat loss versus the temperature difference between the water and its surroundings. If you do this carefully, you will see that the graph is a straight line. This indicates that the rate of heat flow from a warm body is proportional to the difference in temperature between the warm body and its surroundings.

B. *Rate of Heat Flow into Melting Ice*

In this experiment you will measure the time it takes an ice cube to melt in warm water and in cool water. If you use a large amount of water, the temperature around the melting ice will remain nearly constant. If you use identical ice cubes, both of which will be at 32 degrees Fahrenheit, you can be sure that the same amount of heat will flow into both ice cubes in order to melt them. Since the melting rate will depend solely on the rate at which heat flows into the ice, the melting rate can be used to determine the rate of heat flow at known differences in temperature between the water and the ice. The ice will remain at 32 degrees Fahrenheit ($0°$ C) until it is all melted, and the temperature of the water will change very little because there will be so much of it.

Begin by filling a bucket or large container with warm water. Fill a second large container with cool water. Record the temperature of the water in one of the buckets and add one ice cube. Stir the water throughout the melting process to be sure the ice does not become surrounded by its own cold melt water. Measure the time it takes for the ice to melt completely.

Record the temperature in the other bucket and repeat the experiment.

To find the melting rate, divide the total heat used to melt the ice (about 10 Btu's for an average ice cube) by the time it took to melt the ice. Compare the ratio of the melting rates for the two ice cubes with the ratio of the temperature differences between the water and the ice. For instance, if the ice melted in 62-degree water, the temperature difference would be 30 degrees Fahrenheit (62° − 32°). If the other ice cube melted in water at 92 degrees, the temperature difference would be 60 degrees (92° − 30°). The ratio of the temperature differences would be 60/30, or 2 to 1.

You will find that the ratio of the melting rates (which measure the heat flow rates, remember) is very nearly the same as the ratio of the temperature differences between the ice cubes and the water in which they melted.

3. INSULATION—R VALUES

In this experiment you will measure the R value for two styrofoam cups. One will be of single thickness; the other of double thickness. To make a double-thick cup, place one cup inside another. Use scissors to cut off the part of the inner cup that projects above the top of the outer cup. The sides of several other cups can be used to make covers for the two cups. A single layer of siding should cover the top of one cup; a double layer of siding can be used to cover the double-thick cup. See the drawing on the next page.

Which cup do you think will have the larger R value?

To determine the R values, put 100 ml (100 grams,

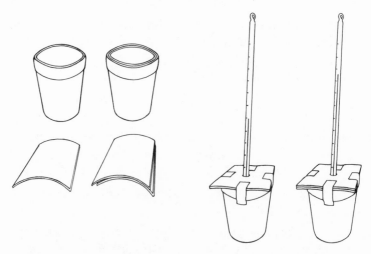

0.22 lb, or 3½ ounces) of hot tap water in each cup.
Tape the covers to the cups and measure their initial
temperatures. The temperatures should be nearly the
same. Leave the cups for ten minutes, but gently swirl
them every minute or so to keep the water temperature
uniform.

After ten minutes (0.17 hour), record the temperatures. How much did the temperature change in each
cup? How much heat, measured in Btu's, was lost from
each cup? What was the temperature of the air around
the cups?

Since R is measured in $\dfrac{\text{square feet}/\text{°F}/\text{hour}}{\text{Btu}}$, you

will need to know the surface area of the cups in square
feet, the time in hours, and the temperature difference
between the inside and outside of the cup. (You can use
the average temperature of the hot water and the air tem-

perature around the cups to determine the temperature difference.)

Now you are ready to calculate the R value for each cup. You can calculate R directly:

$$R = \frac{\text{area} \times [\text{T(inside)} - \text{T(outside)}] \times \text{time}}{\text{heat}}$$

or you can calculate the heat conducted from the container per square foot of area per hour per degree temperature difference and determine the R value from $R = \dfrac{1}{C}$.

What is the R value for each cup? How do they compare?

You may be surprised to find that the ratio of the R values is not 2 to 1. The air layer that clings to the cups' surfaces has an R value too. The effective R value of this air layer is hard to measure, but I estimate it to be about $0.35 \dfrac{\text{square feet/}^\circ\text{F/hour}}{\text{Btu}}$, hence, the R value for each cup will be 0.35 less than the value you determined.

After 0.35 is subtracted from each R value, how do the R values for the two cups compare?

What R value would you assign to a one-inch-thick piece of styrofoam according to your measurements?

Now that you know how to determine the R value for a styrofoam cup, see if you can design experiments to measure the R values for newspaper, fiberglass, metals, plastics, and a thermos.

You can also investigate the effects of placing insulating materials around the sides, over the top, on the bottom, or over the top, and around the sides of a cup of hot water. Where does insulation seem to have the greatest effect in reducing heat losses?

4. ICE CUBE KEEPING

Now that you're an expert on insulation, you might like to run a contest and apply what you have learned. Offer a prize to anyone who can build a better ice cube keeper than you. You'll need some rules, of course. The ice cube keeper must be placed in a warm room, there must be some way to determine when all of the ice has melted, and of course freezers, refrigerators, and extra ice are "illegal."

5. GLASS AND HEAT LOSSES

It should not surprise you that heat flows faster through windows than through walls. If you put one hand on a window on a cold day and your other hand on an adjacent wall, you will feel the difference. The R value for a well-insulated wall is about 14, whereas the R value for a glass window is 1.

Many building codes or guidelines restrict the glassed area of exterior walls to 10 percent of the total floor area and 20 percent of the total exterior wall area. For example, if your house has 1,200 square feet of floor, 1,200 square feet of exterior wall surface, and 120 square feet of glass within the window frames, then glass is 10 percent of the floor area ($120/1200 = 0.10$) and also 10 percent of the exterior wall area.

Do your home and school meet the guidelines suggested above?

What can be done to reduce the amount of heat conducted through existing windows?

6. CHECKING FOR INSULATION IN WALLS

It's usually easy to check the insulation above the ceilings of a house. You simply go into the attic and look. But you can't look into the walls as easily.

Here's one way to tell if the insulation in the walls of a house is adequate. On a cold day, place a thermometer against the inside surface of an exterior wall. After the temperature becomes constant, record the reading and place the thermometer on a chair in the center of the room. Again, record the temperature after it stabilizes. If the wall is more than 5 degrees cooler than the room, the wall is probably not properly insulated.

For the purpose of comparison, place the thermometer against a windowpane and record the temperature after a few minutes. What is the difference in temperature between the window and the center of the room?

7. THERMOSTATS

A thermostat is a switch that is turned on and off by changes in temperature. A thermostat contains a bimetallic strip (strips of two metals, usually iron and brass, that are fastened firmly together). As the temperature in your home decreases, the strip in the thermostat begins to bend. When it has bent enough to touch another piece of metal in the thermostat, an electrical connection is made that turns on your furnace or other heat source. If the thermostat is set for a lower temperature, the bimetallic strip must bend more before electrical contact is made.

You can best see how a thermostat works by making and heating a bimetallic strip. To do this, cut and straighten a strip of thin steel from an ordinary can (commonly called a "tin" can). Then cut a strip of aluminum from a frozen pie or cake pan. Make both strips as long as possible (six to twelve inches). Fasten the two metal strips together with epoxy glue and place the bimetallic strip under a board with a weight on it so that the two metals will become firmly stuck together. If you wish, you can trim the strip after the glue has dried.

Clamp one end of the bimetallic strip to the edge of a chair, stool, or table. Fix a sheet of cardboard beside the other end of the strip and make a mark on the cardboard to indicate the position of the end of the strip. Heat the metal with a burning match or candle. Which way does the bimetallic strip bend? Which metal expands more, aluminum or steel?

8. Detecting Infiltration with a Draftometer

On a brisk windy day in winter you may feel cold air flowing in around a poorly caulked or weather-stripped window or door. With a draftometer you can detect less violent air currents on milder days.

To make a simple draftometer, tape one end of a strip of plastic wrap (6 inches × 3 inches) to a pencil. If you hold the draftometer near a window or door where there is significant infiltration, you will see the strip of plastic "dance" in the moving air. (Be careful not to mistake static electric effects for moving air.)

What can be done to reduce or eliminate the infiltration around doors and windows?

9. BE A HOUSE DOCTOR

Armed with draftometer and thermometer, you can give "physical examinations" to houses and schools. Where is air leaking into your home or school? In which rooms do your thermometers indicate insufficient insulation? Take a look in the attic. Is it well insulated? Is it well ventilated? (The ventilation openings under eaves and at the ends of the attic should be at least $\frac{1}{150}$ the area of the attic floor.)

What can be done to reduce infiltration heat losses? What can be done where there is insufficient insulation? Can heat losses through windows be reduced? What conservation measures can be instituted that would reduce the use of fuel?

ENERGY QUIZ

A. Questions 1 through 8 are based on Tables V and VI.
1. In Chicago, Illinois, which month is coldest?
2. For which month of the heating season will a Chicago home have the lowest fuel bill?
3. The average temperature in February is colder than the average temperature in December. Why

then are there more degree-days in December?

4. If the fuel bill for a Chicago home for April was $70, estimate the bill for March.

5. Where could you live in the United States and never need a heating system?

6. Which city listed in the table is the coldest?

7. Two identical houses are found in Buffalo and Nashville. How will the fuel oil consumption for the two houses compare?

8. The average temperature in Miami is higher than in Honolulu; yet Miami has more degree-days. Explain.

9. A home is kept at 65 degrees during a ten-day period when the average outside temperature is 35 degrees. How many degree-days accumulate? What percent of the fuel could be saved if the temperature were kept at 55 degrees for the same period?

10. How much heat per hour is lost through one square foot of a wall that has an R value of 20 if the temperature inside is 65 degrees when it's 45 outside?

B. True or False

1. Vapor barriers should always be placed on the warm side of the insulation.

2. Heat flows from warm bodies to cooler ones.

3. Failure to use a vapor barrier will increase the R value of the insulation behind the barrier.

4. In a cold climate there should be as few windows as possible on the north side of a house.

5. If you have a limited amount of money, invest it in storm doors rather than storm windows.
6. It is worthwhile to cover windows with plastic sheets if you can't afford storm windows.
7. You can't reduce energy costs by turning down the thermostat at night because it takes more energy than you saved to bring the cool air back up to 65 degrees.
8. A fireplace fire may cause a net loss of energy.
9. If a house has no insulation, the first place to insulate is the attic because warm air rises.
10. Turning off the heat in one-half of a house will cut fuel bills in half.

5

SAVE
THAT GASOLINE

THE WORLD's fleet of 300 million automobiles consumes one of every five barrels of oil produced. In the United States, where more than 40 percent of these cars can be found, automobiles guzzle one-third of the oil and 13 percent of the total energy consumed each year. Automobiles account for 90 percent of the miles traveled by Americans. The average citizen, who drives 10,000 miles annually and purchases 700 gallons of fuel, loves his car so much that he is willing to spend 15 percent of his income on that vehicle. The result: a national fleet of cars traveling more than one trillion miles each year while burning 80 billion gallons of gasoline.

Transportation accounts for 25 percent of the energy this nation uses each year. Table I summarizes the annual energy distribution among the various modes of transportation.

TABLE I

Mode	Percentage of energy consumption	Mode	Percentage of energy consumption
Cars	55	Airplanes	9.9
Trucks	20	Railroads:	
Buses	1	Freight	3.5
Other (mopeds,		Passenger	0.1
motorcycles,		Ships	4.8
etc.	½	Fuel pipelines	5.2
Total highway			
transportation	76½	Total	100.0

GASOLINE AND DIESEL ENGINES

An automobile carburetor is designed to change liquid gasoline into a flammable vapor by mixing it with air in a weight ratio of sixteen parts air to one part fuel. The gasoline vapor and air pass through the intake valve to the cylinder where a moving piston compresses the air-fuel mixture to one-eighth its volume. As the piston reaches the peak of its stroke in "squeezing" the gas, an electric potential of 20,000 to 30,000 volts across the spark plug gap generates a spark that ignites the flammable mixture. The burning gas expands, driving the piston back and transferring kinetic energy to the crankshaft. During the next upstroke of the piston, as fuel burns in another cylinder, the exhaust valve opens and the waste gases produced are expelled to the exhaust manifold and muffler.

The crankshaft converts the up-and-down motion of the pistons to the rotary motion of the flywheel, drive shaft, gears, axles, and wheels.

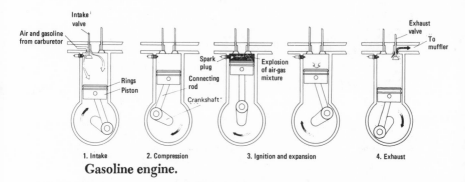

1. Intake   2. Compression   3. Ignition and expansion   4. Exhaust
Gasoline engine.

The efficiency of gasoline engines can be improved by higher compression ratios, but the higher temperatures produced by high compression create a need for additives to reduce the fuel's tendency to explode (knock) rather than burn. Improved design of the combustion chambers to ensure a thorough mixing of fuel and air and, thereby, complete combustion will improve efficiency and reduce the oxides of nitrogen and sulfur as well as carbon monoxide in the polluting exhaust fumes, but it will be difficult to obtain efficiencies greater than 15 percent.

A more efficient (25 percent), less polluting, internal combustion engine that powers many trucks and is becoming more common in cars is the diesel. Diesel engines use a heavier, less volatile fuel that stores more energy per gallons and requires none of the tetraethyl lead used to reduce the knocking tendency of gasoline.

Unlike a gasoline engine, the diesel has no spark plugs. A compression ratio at least twice the 8 to 1 ratio characteristic of gasoline engines raises the fuel-air mixture to a temperature above its ignition point. At the top of the compression stroke, a small amount of fuel is sprayed into the very hot compressed air. The burning mixture drives the piston through a long power stroke that acts on the crankshaft for a longer period than in the gasoline engine. Because the air drawn into the cylinder exceeds the volume needed for complete combustion, the polluting oxides so characteristic of gasoline engines are reduced in quantity.

Diesels are not without fault, however. Although they are more efficient and effective in reducing pollution, they are heavier, more expensive to build, and noisier.

IMPROVING YOUR MILEAGE

The ascending price of gasoline has forced car owners to seek more efficient, economical automobiles. A vehicle's fuel consumption is a function of its weight, engine size, and condition of repair as well as its driving conditions and the operator's skill. A lightweight car with a small engine can travel much farther per gallon of fuel than a heavier model; consequently, small car sales have increased while larger models remain in showrooms and lots. American automobile companies have suffered severe losses because they failed to anticipate consumer reaction to soaring fuel prices.

In an effort to force manufacturers to reduce the production of gas guzzlers, Congress, in the Energy Policy and Conservation Act of 1975 (EPCA), required companies to improve the efficiency of their cars according to the schedule shown in Table II. These EPCA standards are "fleet average" values. The average fuel economy of all cars manufactured by a company in any given year must meet the minimum EPCA standards.

TABLE II

MODEL YEAR	"FLEET AVERAGE" ECONOMY STANDARD (MPG)	MODEL YEAR	"FLEET AVERAGE" ECONOMY STANDARD (MPG)
1978	18.0	1982	24.0
1979	19.0	1983	26.0
1980	20.0	1984	27.0
1981	22.0	1985	27.5

The Energy Act of 1978 imposes a tax on any car manufacturer whose "fleet average" fails to meet EPCA standards for any given year.

Recent projections from the trends in sales indicate that 80 percent of new car sales in 1985 will be compacts or subcompacts. (In 1958 about 90 percent of all new cars were standard size.) If these predictions hold up, gasoline consumption by 1985 will be 26 percent less than would have been anticipated two years ago. At this time (spring 1980) gasoline consumption is nearly 10 percent less than it was one year ago. Both the price of gasoline and the switch to smaller cars have reduced fuel consumption significantly.

New small-car fuel economy, higher fuel costs, increased mass transit facilities in cities, more economical planning by airlines, and strict enforcement of the national 55 mph speed limit to reduce the higher fuel consumption (and death rate) associated with speeding could reduce fuel consumption by 2.5 million barrels per day below earlier predictions for 1985.

Here are a number of good driving habits that can conserve energy, reduce driving costs, and improve mileage.

- Don't wait for an engine to warm up. Gasoline is burned while the engine idles. Set the car in motion after a ten-second wait, but drive slowly until the engine reaches its normal operating temperature.
- Don't "rev" a car engine. Every time you press the accelerator more fuel is burned for no reason.
- Accelerate slowly. Jackrabbit starts require high power; that is, the engine must do a lot of work in a short period of time. The work can't be done unless there is energy transferred from the chemical energy in the gasoline to the kinetic energy of the wheels. To do this rapidly requires an increase in the fuel consumed per unit time. Further, to apply the large forces needed for rapid acceleration, the car must be in a low gear ratio where the energy transfer is not efficient.
- Anticipate stops and slow down gradually. Rapid stops, like sudden starts, waste fuel.
- Drive at a steady, moderate speed to improve gas mileage.
- Use a car's air conditioner only when absolutely necessary. Some of the car's energy is used to power the air conditioner.
- Remove excess weight from your car. It takes energy to move weight.
- Park your car in a central area and leave it when you reach your general destination. Short trips, such as

from store to store, in a shopping center are fuel wasters. Park in a central location and carry or wheel packages to the car.

- Use snow tires for as short a time as possible. Additional energy is used to overcome the extra friction created by snow tires on bare roads.
- Have your engine tuned frequently to maintain top mileage performance.
- To avoid fouled spark plugs and poor gas mileage, use fuel with the proper octane rating for your car.
- Have battery fluid level, belts, hoses, brakes, spark plugs, automatic choke, and thermostat checked frequently.
- Don't allow service station attendants to fill your fuel tank to the brim. Gasoline may leak out later if you park on a hill.
- Walk or bike to stores if you need only a few items. Bargain shopping that requires driving for small items saves pennies but costs dollars for fuel.
- Vacation near home or at a location where a car is unnecessary.
- Use public transportation whenever possible.
- Encourage and help establish car or van pools for travel to work, shopping, and social activities.
- Walk, bike, or ride a bus to school; don't use an automobile.
- Plan shopping trips to reduce automobile use: (1) Keep a list of errands so that all of them can be accomplished in one trip. (2) Take the most direct, efficient route. (3) Be sure that the most economical car is driven as much as possible if your family has several automobiles. (4) Buy enough groceries to last

a week or more. (5) Arrange to have several family members go for routine medical and dental checkups at the same time.

EXPERIMENTS

1. MILES PER GALLON

How far can your family car go on one gallon of gasoline? One way to find out is to drain the fuel tank, add one gallon of gasoline (or diesel fuel), note the odometer reading, drive the car until it stops, and again read the odometer. The difference in odometer readings will tell you how far you've traveled on one gallon of fuel. Since you might run out of fuel in rush hour traffic or on a deserted highway, there are more practical methods for measuring miles per gallon.

To obtain an estimate of your car's fuel economy, record the odometer reading when the tank is being filled at a service station. When the tank is nearly empty, stop for fuel and again record the odometer reading *and* the number of gallons needed to refill the tank. Dividing the number of miles traveled since the tank was last filled by the gallons of fuel added to the tank will give you the miles traveled per gallon. Of course, if the car is driven in stop-and-go traffic throughout the time between fuel stops, your value for miles per gallon might be rather disappointing. On the other hand, if all the fuel is used while driving along flat, straight roads at 35 mph, the miles per gallon calculated will be much greater than you could expect for normal driving conditions.

To obtain a more accurate measure of your car's mileage, keep records of the fuel consumed over several thousand miles of driving. You should also record any unusual conditions such as winter driving on ice or snow, travel over mountains, through heavy traffic, on gravel roads, and so forth.

2. TIRE PRESSURE AND EFFICIENCY

To see how the pressure of the air in tires affects the efficiency of a vehicle, try this experiment.

First, find a place where you can use your bike to coast down a hill and then along a level or slightly uphill path. Try this with your bicycle tires inflated to the pressure called for on the tires. Mark the point where your bike will coast no farther. Repeat the run, but this time reduce the pressure in the tires by half. How far do you coast now? Try the same run again with the tires over-inflated.

On the basis of your experiment, do you think an automobile operates more efficiently on underinflated, properly inflated, or overinflated tires? Why?

Why do you think some people insist on having the tires on their cars underinflated?

3. WEIGHT AND EFFICIENCY

It takes energy to do work. Work is force times distance ($F \times D = W$). If you exert a force of 10 pounds through a distance of 5 feet, you do 50 foot-pounds of work.

With a spring balance, measure the force required

to pull an empty wagon along a floor, road, or sidewalk at a steady speed. Calculate the work you do in moving the wagon through a distance of 10 feet. Now, ask someone to sit in the wagon. How much force do you have to exert to pull the loaded wagon over the same path at a steady speed? How much work would be required to move the loaded wagon through a distance of 10 feet?

On the basis of your experiment, why do you think more and more people in the United States are driving small cars?

AUTOMOBILE ECONOMY QUIZ

A. True or False
   1. To conserve fuel you should always let a car's engine warm up before you begin to drive.
   2. Driving at 40 mph uses less fuel than driving at 60 mph.
   3. Driving in second gear consumes less fuel per mile than driving in third or fourth gear.
   4. Jackrabbit starts waste fuel.
   5. Air-conditioned cars are more economical to drive than those without this convenience.
   6. Large, heavy cars develop more momentum and therefore are more economical to drive than smaller cars.
   7. Proper maintenance can improve a car's mpg ratio.
   8. People should ride buses because a bus travels farther on a gallon of gas than an automobile does.
   9. People should ride buses because buses can ob-

tain more passenger miles per gallon than cars can.

10. Careful planning of shopping trips is a waste of time because the distance to the stores does not change.

B. Here is the mileage-fuel record of a conservation-minded driver who kept careful check on the mpg ratio.

Date	Odometer Reading (miles)	Fuel Added (gallons)	Notes
9/20	00340	10.2	First fuel added to new car
9/27	00675	10.1	
10/1	00920	8.0	
10/10	01303	12.3	
10/16	01613	10.8	After trip through mountains
10/21	01954	11.1	
10/30	02304	10.1	After trip on express highway
11/4	02626	9.8	
11/12	02898	8.5	
11/20	03245	10.0	
11/28	03578	12.0	After long trip on snow
12/5	03888	10.2	
12/12	04212	11.1	During very cold weather
12/20	04548	11.7	During more cold weather

1. What was this car's mpg ratio over the three months from September to December?

2. What factors seem to reduce this car's mpg ratio?

3. What factors seem to increase this car's mpg ratio?

CHAPTER

6

OTHER WAYS
TO CONSERVE ENERGY

INSULATION, WEATHERIZATION, conservation practices with electrical appliances, reduced driving, and proper driving habits can all reduce energy use. But there is plenty more that can be done to improve energy efficiency in your home and school.

MONEY UP THE CHIMNEY

An inefficient oil or gas furnace will send money up the chimney; it's well worth the cost to have a representative from your oil or gas company check and clean your furnace annually.

A heating system that burns fuel oil, gas, coal, or wood consists of four principal parts: burner, furnace or boiler, distributing system, and chimney. Heat from the burning fuel is absorbed by the furnace or boiler that transfers heat to the distributing system. The warm air or water is circulated in the building by air ducts and registers or water pipes and radiators. As the hot gases from the burner pass through the furnace, not all the heat is transferred to circulating air or water; some ascends the chimney, thereby reducing the efficiency of the system.

When examining your heating system, the furnace "doctor" carries out a number of tests after cleaning the furnace and drilling a small hole (if one is not there already) in the flue pipe or stack. (The hole is plugged later with a sheet-metal screw.) One test measures the percent of carbon dioxide (CO_2) in the gas leaving the furnace. When fuels burn, they combine with oxygen to form carbon dioxide, water vapor, and other gases such as sulfur dioxide. A low percentage CO_2 reading indicates that the fuel is not burning completely. Although 6 to 8 percent CO_2 is good for older units, new units should produce at least 10 percent. If the reading is low, the air-fuel mixture can be adjusted to obtain better combustion.

Another test measures the exhaust gas temperature. Low stack temperatures indicate an efficient heat transfer in the furnace, but temperatures below 300 degrees Fahrenheit could produce corrosion when sulfur dioxide and water vapor condense to produce sulfuric acid. Normally, stack gas temperatures should be 300 to 500 degrees.

A smoke test is performed by slipping a patch of special material into the stack for a fixed time. The patch is then compared with a set of standards. A rating of zero to one is good; sootier patches indicate incomplete combustion and the possible buildup of a sooty deposit in the furnace that prevents efficient transfer of heat.

A draft test of 0.04 to 0.06 indicates that there is enough air entering the burner to burn the fuel completely. An excess of air may cool the gases and reduce furnace efficiency.

After the initial tests, the furnace technician makes a number of adjustments and tests until a clean fire (minimal smoke) is produced with minimum draft and maximum CO_2. A chart that integrates these readings enables the technician to determine the efficiency of your furnace.

The maximum efficiency expected of a furnace is 88 percent, but many drop to 50 percent or worse unless "tuned" periodically. Table I indicates the savings per $100 of fuel cost as furnace efficiency is improved.

TABLE I

EFFICIENCY BEFORE CHECKUP	EFFICIENCY AFTER CHECKUP			
	70%	75%	80%	85%
50%	$28.60	$33.00	$37.50	$41.20
60%	14.30	20.00	25.00	29.40
70%	—	6.70	12.50	17.60
80%	—	—	—	5.90

Here are a number of improvements or changes that can be made in a heating system to improve its overall efficiency.

- Installing an automatic damper will prevent hot air from escaping up the chimney when the burner unit is off. These devices are most effective on old, inefficient furnaces that are oversized—that is, furnaces in which the burner unit is off much of the time. Savings of 2 to 13 percent can be realized with automatic dampers.

- Replacing pilot lights on old gas furnaces with an electric ignition system will improve seasonal efficiency by 10 percent.
- A sealed combustion unit that uses cold outside air rather than warm inside air for combustion can improve furnace efficiency by 4 to 20 percent.
- Proper sizing of a furnace can reduce fuel costs 5 to 10 percent. Many furnaces are too large; they quickly bring a building to the temperature called for by the thermostat and are then off for long periods of time. The entire furnace has to be warmed each time the burner goes on. Changing to a smaller burner nozzle will provide less fuel per unit time causing the burner to be on for longer periods. If properly sized, your heating system should be on continuously during the coldest days of winter.
- Installation of a flame-retention head burner will improve an old furnace that can't be adjusted to burn at better than 70 percent efficiency. Although these units are expensive, they mix fuel and air in just the right proportions to give slower, more efficient combustion.
- Stack heat reclaimers where stack temperatures exceed 450 degrees Fahrenheit will recover heat that would normally escape through the chimney.

There are also a number of things you can do that involve no major changes but will reduce heating costs.

- Set fan switches to go off at 75 degrees and come on at 90 degrees rather than the usual 100 and 125. This resetting can result in fuel savings of 6 to 10 percent.
- Insulate heat pipes or ducts to reduce heat losses, particularly pipes or ducts that carry heat through long sections of unheated space.
- Change or clean the air filters on hot air furnaces frequently in the winter.
- Clean radiators or baseboard heaters, and be sure convection currents are not obstructed by furniture or draperies. Aluminum foil behind radiators will reflect more heat into the room.
- If you have hot water heat, ask your furnace repairman to show you where to drain water from the furnace. By removing a bucket of water from the pipes each month, you can improve circulation by getting rid of unwanted rust and other particles. You should also know where the valves are that enable you to add water to the boiler when the gauge indicates it's needed.
- Bleed air from radiators. Air gets into the pipes and rises to the tops of the radiators where it prevents hot water from entering. A valve at the top of the radiator can be opened every six months. Hold a pail under the valve until water comes out. Be careful—the water may be hot.

WOOD STOVES

During this nation's first hundred years, wood was the primary source of heat. In many parts of the country, where laws allow it, people are returning to the woods for fuel.

Before you rush out to buy a stove, there are a few things you should know. First, you may have to cut and split your own wood—a time-consuming process, and certainly you will have to build, stoke, and check the fire, learn to operate dampers and vents, carry wood, and remove ashes. A wood stove demands attention; it won't go on and off automatically.

If, realizing the work involved, your family decides to buy a stove, be sure to obtain a permit from your local building inspector. The permit will specify certain requirements. Most towns demand that the stove be three feet from any walls, with its legs on a noncombustible surface such as a hearth or a stove board.

Most good stoves are designed to burn wood slowly. As a result, creosote, a flammable oily liquid formed by the condensation of gases created when wood burns, can accumulate in a chimney and ignite. To avoid such chimney fires, you should burn at least one hot fire each day to vaporize creosote. You can reduce creosote buildup by burning hard woods such as oak, hickory, and maple, while avoiding the softer pine, fir, and spruce. Monthly chimney cleanings will also reduce chimney fire danger. If the stove is connected to an existing chimney, the stovepipe should be as short and straight as possible. Many families install a separate, stainless steel stack when they purchase a wood stove.

To start a stove fire, place wood on kindling and open the damper before lighting the kindling with paper. (Don't use charcoal lighter fluid to start a wood fire!)

Once the fire starts, close the stove door, shut down the damper, and regulate the vents to control the temperature. Opening the vents will admit more air and produce a hotter, more rapidly burning fire. Closing the vents, which many people do just before going to bed, will reduce the burning rate.

To restart a fire, first open the damper so that smoke will ascend the chimney and not billow out into the room. Pull hot coals to the front of the stove and stack wood on them. When the wood is burning, close the door and damper.

Ashes must be removed from the stove periodically. Scoop them into a *metal* bucket. (Any hot coals might ignite a nonmetallic container.)

FUEL FOR STOVES

We could harvest 250 million cords of fuel from America's forests each year with no danger of depleting our supply of wood. Since five cords* will heat the average home for one year, fifty million homes—approximately 80 percent of our dwellings—could be heated by wood.

Denser woods (hard woods) are preferred for heating. Less dense wood (soft wood) is generally used for pulp and lumber. Table II gives the energy content, in millions of Btu's per cord, of some common woods, as well as an estimate of what it costs to obtain the same heat from fuel oil.

---

* A cord of wood is a stack 4 ft. × 4 ft. × 8 ft. (128 cu. ft.).

TABLE II

Type of wood	Average density (lbs per cord)	Energy content (m btu)	Cost (in dollars) to obtain same heat from fuel oil at $1 per gal. (assumes stove efficiency = 50%, oil furnace = 75%)
Hickory	4,400	31.2	149
White oak	4,400	31.2	149
Sugar maple	4,100	29.1	139
American beech	4,000	28.4	135
Red oak	3,900	27.6	131
Yellow birch	3,800	27.0	129
White ash	3,700	26.3	125
American elm	3,400	24.1	115
Red maple	3,400	24.1	115
Paper birch	3,400	24.1	115
Black cherry	3,300	23.4	111
Douglas fir	2,900	21.5	102
Eastern white pine	2,200	15.8	75

The price in the right-hand column is the break-even price for a cord of wood. If wood costs more per cord, you'd do better to use fuel oil. The price does *not* include your time in tending the stove; however, if you enjoy cutting wood, you can probably buy it for well under $100 per cord if you supply the labor.

AIR CONDITIONERS AND HEAT PUMPS

Air conditioners are designed to cool, dehumidify, and filter air. A complete system will heat air as well as cool it. The size of a system varies with the building and its needs. Many families have air conditioners that cool one room. They are commonly seen projecting from the lower portions of windows.

Most air conditioners are similar to refrigerators. A refrigerant (Freon, ammonia, sulfur dioxide, or another liquid that boils at a low temperature) under pressure is allowed to expand as it absorbs heat and boils. The fluid is then compressed outside the space being cooled (outside the window in the case of a small one-room air conditioner). Under pressure, the refrigerant becomes hotter than its surroundings and heat flows out of the fluid before it returns to the cooler environment where it will absorb more heat.

Heat pumps are air conditioners that can be run backwards during the winter to supply heat. When it's cold outside, the compressor is used to heat internal air space. In warm weather the expansion portion of the heat pump absorbs heat from air inside the building.

Heat pumps are generally most economical in moderate climates where heating and cooling energy demands are equal. If heating demands greatly exceed cooling requirements, the heat pump must have a powerful compressor to meet the heating needs, a supplementary heating system, or a large heat-storage system such as a pond or pool.

To keep an air conditioner working efficiently, it should be serviced annually; air filters should be cleaned or replaced every month, and you should try to shade the compressor to improve heat transfer to the outside air. Be sure, however, that you do not obstruct the air flow around the compressor.

INSULATED SHADES

Windows are energy wasters. With an R value of 1 and the possibility of infiltration around the frames, windows generally account for a large portion of the heat lost from a building.

Drawing shades and draperies can reduce heat losses by 15 percent, but new insulated shades over double glazed windows can produce an R value of 17, making windows nearly as resistant to heat flow as well insulated walls. Early models were built on standard shade rollers and were clumsy to operate. A more recent design has a bead chain and sprocket arrangement that is easier to operate. These shades have five layers and sell for about $3.00 per square foot, but they must be installed carefully within a frame to ensure a good seal around the window.

Curtain Wall with an R value of 9 to 12 is a large multilayered shade that can be used on picture windows and sliding glass doors. It costs about $4.50 per square foot and can be made fully automatic when electrically connected to a thermostat.

Window Quilt sells for about $3.75 per square foot. It is a thick roll-up shade with an R value of 5. Its sides slide within plastic tracks attached to the window frame.

If you're a do-it-yourselfer who doesn't mind daily chores, you can buy styrofoam sheets and cut them to fit windows, picture windows, and glass doors. Eye hooks and long rubber bands can be used to hold these insulating sheets in place at night.

LANDSCAPING

If your family is building a home, try to position the house so that it is sheltered from north winds. A windbreak on the north and northwest sides of your house should be placed at a distance from the building that is equal to twice the height of the house. These evergreen trees and bushes could reduce heating bills by 20 percent or more. Deciduous trees on the south and west sides of your house can provide summer shade while allowing the sunlight's warmth to enter your home in winter.

Leaves raked from the lawn in autumn can be bagged and placed around the foundation. They provide excellent insulation against winter's bitter cold.

EXPERIMENTS

1. COOLING BY EVAPORATION

If you stand in a breeze after taking a swim, you may begin to shiver even on a hot day. You'll feel cooler in front of a fan or near an open window in a moving car or bus. Is moving air cooler than still air?

To find out, hold a thermometer by a string in front of a fan and then in still air (turn off the fan). Does the temperature rise when you turn off the fan?

Soak a piece of thin cloth in lukewarm water. Squeeze out excess water and fasten the damp cloth to the thermometer bulb. Note the temperature when the thermometer liquid stops moving. Hold the thermometer in front of a fan or swing it through the air. What happens to the temperature?

Eventually the cloth dries because water evaporates into the air. The fastest moving (hottest) molecules escape from the liquid and become a gas (water vapor). As these faster moving molecules escape, the ones that remain are the slower moving (cooler) molecules. So you see, evaporation of water has a cooling effect. That's why you may be chilled if you stand in a wind while wet. It also explains why perspiring helps to keep your body cool.

To see which of several liquids evaporates fastest, make some liquid streaks with your finger on a piece of foil. Try water, rubbing alcohol, and cooking oil. Which liquid evaporates fastest? Slowest? Which liquid do you think will produce the greatest cooling effect?

To test your prediction, place a little of each liquid on the back of your hand. Which liquid makes your hand feel coolest?

Design your own experiments to see how each of the following affects the evaporation of water: temperature, wind, humidity (moisture in the air), surface area. Of course, you will need at least two setups for each experiment so that you can compare results. For example, to test the effect of temperature, you will need equal amounts of water in identical containers at different temperatures.

2. Cooling and Rate of Evaporation

To see how the rate of evaporation affects the cooling due to evaporation, try this. Pour equal amounts of hot tap water into each of three identical aluminum pie pans. Place the pans on folded newspaper or cardboard insulating sheets. The pans should be several feet apart. Submerge a thermometer in each pan. To keep the evaporation rate near zero in one pan, add a few drops of cooking oil to coat the water. Leave the second pan undisturbed. Increase the rate of evaporation in the third pan by using a fan to blow air across it.

Record the temperature in each pan at one-minute intervals for a few minutes. In which pan does the water cool fastest?

3. Humidity

Just as sugar dissolves in water, so water dissolves in air. We can't see water vapor in the air, but the condensation on a cold window reveals its presence.

The solubility of sugar increases in warm water; the solubility of water in air increases as the air is heated. But there is a limit to the amount of sugar that will dissolve in a quart of water at any particular temperature. When this limit is reached, we say the solution is *saturated*. Water vapor also has a saturation point in air at a particular temperature. Table III gives these saturation points for several temperatures.

TABLE III

MASS OF WATER VAPOR IN ONE CUBIC METER OF AIR WHEN
SATURATED

TEMPERATURE	WATER VAPOR (GRAMS)	TEMPERATURE	WATER VAPOR (GRAMS)
32°F (0°C)	4.8	68°F (20°C)	17.1
41°F (5°C)	6.8	77°F (25°C)	22.8
50°F (10°C)	9.3	86°F (30°C)	30.0
59°F (15°C)	12.7	95°F (35°C)	39.2

The quantity of water vapor dissolved in one cubic meter of air is called the *absolute humidity*. It measures the dampness of the air.

You can measure absolute humidity by determining the dew point. For example, suppose the air in a room is 77 degrees Fahrenheit (25° C), and you slowly lower the temperature of some warm water in a shiny metal can by adding small pieces of ice to the water as you stir it. At the dew point, moisture (dew) can be seen condensing on the cold can because the air in contact with the can is saturated with vapor, and any excess will condense. If the moisture is first seen when the can is at 50 degrees Fahrenheit (10° C), you know from Table III that the air holds 9.3 grams per cubic meter—the saturation point for air at 50 degrees (10° C). Since the air not in contact with the cup is at 77 degrees Fahrenheit (25° C), the air could hold as much as 22.8 grams per cubic meter. It is far from saturated.

In fact, it holds 9.3 of the 22.8 grams it could hold. It has only 41 percent (9.3/22.8) of the moisture it could hold. This ratio—the water vapor present in the air compared to the total vapor the air could hold if saturated—is called the *relative humidity*.

Use the dew point method to find the *absolute* humidity of the air inside and outside your home or school. What is the *relative* humidity both inside and outside?

Repeat this experiment on clear, cloudy, and rainy days. Repeat it at different times of the year. What changes in absolute and relative humidity do you find?

Meteorologists often determine relative humidity with a sling psychrometer. See if you can find out what this device is. You could even build one and see how easy it is to measure relative humidity with this instrument.

ANOTHER ENERGY QUIZ

Choose the correct answer from the four given.

1. A repairman should service a furnace (a) daily, (b) weekly, (c) monthly, (d) annually.
2. An efficient furnace, when tested, will show a minimum draft for maximum CO_2 and minimal (a) oxygen, (b) smoke, (c) temperature, (d) heat.
3. An oversized furnace will (a) never go on, (b) be on all the time, (c) never go off, (d) go on and off frequently.
4. One serious danger associated with wood stoves is (a) chimney fires, (b) suffocation from dust, (c) poisonous ashes, (d) dry air.
5. Heat pumps can be used to (a) pump heat from one house to another, (b) both heat and cool a house, (c) heat water pumps, (d) cook food.

6. One way to reduce heat losses from a house is to (a) add more windows, (b) add more doors, (c) install insulated shades over windows, (d) install fans.

7. On which side of a house is a windbreak generally most effective in reducing heat losses? (a) north, (b) east, (c) south, (d) west.

8. To best reduce summer cooling, you should have deciduous trees on which side(s) of your house? (a) north, (b) east and west, (c) south, east, and west, (d) north and south.

9. Which of the following liquids will have the greatest cooling effect when it evaporates? (a) water, (b) salt water, (c) cooking oil, (d) alcohol.

10. The dew point is 50 degrees Fahrenheit (10° C) on a day when the temperature is 68 degrees Fahrenheit (20° C). The relative humidity is (a) 54%, (b) 9.3g, (c) 17.1g, (d) (17.1 − 9.3)%.

SOLAR:
ENERGY OF THE PAST AND FUTURE

STAND IN front of a sunny window while you're wearing a dark shirt. You can feel the heat soaking into the dark cloth. This observation is not new; it was made thousands of years ago. Newspaper accounts of the energy crisis and its solutions imply that solar energy is a new idea. Anthropologists will tell you that people used the sun to heat their dwellings long before written history began. The Pueblo Indians of the southwestern United States lived within thick adobe walls that absorbed the sun's energy and then radiated the stored heat to the occupants at night as the temperature of the desert air plummeted.

Early humans in the northern hemisphere often lived in caves with south-facing openings that allowed the low winter sun to warm the interior while providing shade from the hot, higher, summer sun.

Architects and builders, guided by the abundance of inexpensive energy during the first seven decades of the twentieth century, tended to ignore the sun as they planned and built houses, apartments, offices, and schools. The ancient art of orienting buildings to utilize the sun's warmth was forgotten, but the skyrocketing cost of energy since 1973 has fostered a rebirth of solar energy use in America.

Solar energy is free. Unlike nuclear energy, most of the sun's harmful radiation is removed by the atmosphere. Solar energy produces none of the polluting gases or particles associated with coal or oil, and we do not have to tear the earth open to find it. Sunlight falls upon every nation of the world and cannot be controlled or hoarded by nations or terrorists.

The south side of this building has no windows in the living area.

The rate at which energy flows from the sun to our atmosphere is 7.15 Btu's/sq. ft./ min., 1,353 watts/sq. meter/min., or 1.95 cal./sq. cm./min. In one year 2.6 $\times$ 10^{21} Btu's or 7.5 $\times$ 10^{17} kwh of solar energy strikes our atmosphere. Of that, 50 percent is reflected back into space by the atmosphere. Another 15 percent is reflected by the earth's surface. About 5 percent is absorbed by soil, nearly 2 percent by marine vegetation, 0.2 percent by land plants, and the rest by the water in oceans and lakes. If we could capture just 0.01 percent of the solar energy that comes to the earth's atmosphere, or 0.2 percent of the sunlight that illuminates the continents of this globe, we could meet the entire world's energy demand.

Unfortunately, solar energy is diffuse and difficult to concentrate into a usable form; however, there are ways to capture and use at least a portion of this free, available energy source.

SOLAR HEATING OF WATER

Solar energy can reduce your hot water bill by 80 percent *if* you can install solar collectors. What are the ifs? First, solar collectors must face the sun. Do you have a roof that faces south? Second, can you afford them? Solar collectors are expensive and the pay-back time may be ten years or more; however, your family can receive credit for a sizable chunk of the cost at income tax time.

But what is a solar collector and how does it work? A collector consists of an absorber surface, a glazing of glass or plastic covering the absorber plate, and a fluid, usually water, or air, that transfers heat from the collector to a hot water storage tank.

The absorber plate is usually copper, aluminum, or steel. It is painted black to absorb light and convert it to heat. The collector in the drawing has a water-carrying copper tube fastened to the absorber plate. Water systems use metal absorber plates to transfer heat because metal conducts heat uniformly from the warm plate to the cool water.

There are usually two transparent covers about ¾ inch apart with the bottom plate about an inch above the absorber. About 80 percent of the short-wave, visible light is transmitted through the covers to the absorber, where the light is converted to heat. The covers produce a greenhouse effect that traps heat in the collector. Glass will transmit the entering visible light, but it is opaque to the long-wave (infrared) radiation emitted by the warm absorber. The greenhouse effect is a familiar one. If you

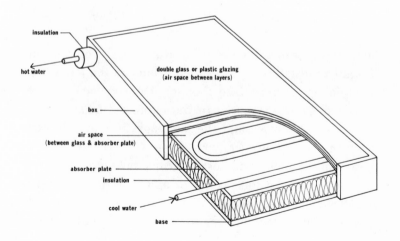

This home uses the hot air generated in the rooftop collectors shown here to produce 80 percent of the family's hot water.

get into a car that has been parked in the sun for a while, even though it's cool outside, the inside of the car will be hot.

There are various collector designs. In some, water flows along passages built into preformed aluminum plates that are fused together. These require no welding or clamping and transfer heat very well. The Thomason system uses corrugated aluminum sheets. Water trickling down the corrugations is heated before it collects in a trough and is gravity-fed into a storage tank from which water is pumped back to the top of the collector. Some absorber plates are coated with selective surfaces that are excellent absorbers and poor emitters of radiation.

In collectors that use air as the absorber fluid, the absorber can be made of a variety of materials: black cotton screening, glass painted black, metal scraps on plywood, metal lath, fiberglass mesh, cloth or paper, and metal. With air, the idea is to create turbulent flow to prevent the buildup of an insulating air layer on the absorber. Fins, "Vees," and rough absorber surfaces are used to produce the turbulence needed for the air stream to absorb 150 to 300 Btu's per hour per square foot of collector surface. The turbulent flow essential for air-type collectors requires a more powerful motor than systems that use water. Further, because of the specific heat (0.24) and density (0.075 lb./cu. ft.) of air, 3,500 cubic feet are needed to absorb as much heat as one cubic foot of water.

Although electrical costs to pump air exceed those for water-type collectors, air has definite advantages as a heat transfer fluid: freezing temperatures cause no concern, the plumbing is inexpensive, there are none of the corrosion problems that plague liquid systems, and the system is easy to repair and maintain.

The efficient operation of a solar heated hot water system requires that you take certain precautions:

- Reduce heat losses by insulating backs and sides of collectors.
- Adjust the flow rate of the fluid to keep the temperature difference between the absorber and its surrounding air as small as possible.

- Collectors must face within a few degrees of true south.
- Ideally collectors should be perpendicular to the sun's rays. A general rule is: Tilt collectors 15 degrees greater than the latitude of the site.
- Install enough collector surface to provide a reasonable fraction of your hot water needs. If you're getting 50 percent from solar, doubling the collector surface will not give you 100 percent solar hot water because many collectors together seldom work at full capacity, and the possibility of a week of rain makes a backup system essential anyway.
- The low cost of solar water heaters, the ease of combining them with a present system, the daily need for hot water, and the ability of a contractor to easily size solar heating for a household make installation of a solar system economically feasible for many homes. Some relatively simple solar water heaters can be built as a do-it-yourself project. You can find books on the subject in your library or bookstore.

SPACE HEATING

Heating your home with solar collectors is far more expensive than heating water. It requires a large water tank or stone bin to store heat for nights and rainy days. Two systems, one using water and the other using air to transfer heat, are seen in the next drawing. Table I gives information about collector areas and storage tank (water) volumes to solar-heat a 1,500-square-foot house in various parts of the country.

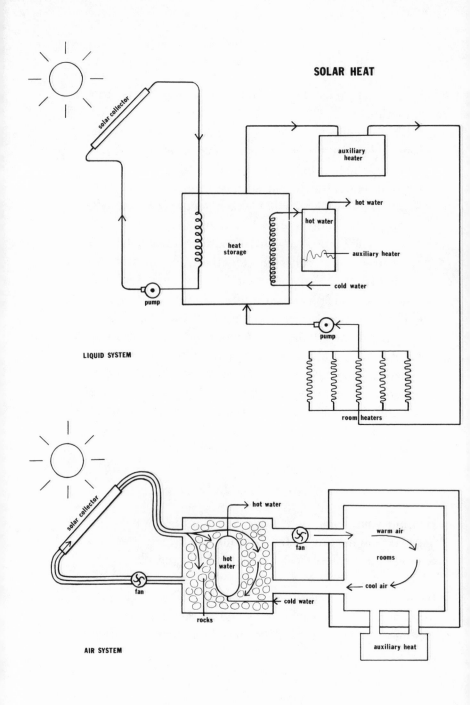

SOLAR HEAT

LIQUID SYSTEM

solar collector

pump

heat storage

hot water

auxiliary heater

hot water

auxiliary heater

cold water

pump

room heaters

AIR SYSTEM

solar collector

fan

hot water

hot water

rocks

cold water

fan

warm air

rooms

cool air

auxiliary heat

This home uses air to transfer heat from the collectors that cover the south side of the roof to a stone storage bin in the basement. The builder expects to obtain 70 percent of his hot water and heat from solar energy.

TABLE I

AREA OF U.S.	COLLECTOR AREA (SQ. FT.)	STORAGE TANK (GAL.)
Northeast	800	1,500
Mid-Atlantic	500	750
Midwest	800	1,500
South and Southwest	200	300
Western Rockies	750	1,500
Northwest	500	800
Middle United States	600	1,000

Rock storage beds for air systems require 100 to 400 pounds of one- to four-inch rocks per square foot of collector.

This solar home produces 90 percent of its heat and hot water from the thirty-three solar panels on the roof. A 1,200-gallon insulated tank stores hot water in the basement. Considerable heat is obtained through the greenhouse on the east side and the many windows on the south side, which are covered at night with insulated draperies. Electric baseboard heat is used as the backup system.

All solar systems require a backup heating system. A wood stove may serve as a backup in rural areas. In most of California and Florida portable electric heaters are sufficient, but in most of the country a full-size gas, oil, or electric system is required.

PASSIVE SOLAR HEATING

Systems that use collectors, pumps, and fluid to move the absorbed solar energy around are called active solar systems. There are other more natural, less mechanical methods to solar-heat buildings. These natural methods are called passive solar systems.

Effective passive solar systems must (1) let solar energy into the building, (2) absorb and store the energy, (3) trap the energy so that it can't escape.

Passive solar homes must have plenty of south-facing windows to let the winter sunshine in. You can keep records to show that more sunlight falls directly on south windows in the winter than in the summer. In fact, with proper roof overhang, summer's high-in-the-sky sun will never illuminate south-facing windows.

ABSORBING AND STORING SOLAR ENERGY

If you build a house with plenty of windows on the south side, it may be so hot inside by noon on a bright January day that you'll want to open a window to let excess heat out; yet, after sunset, heat losses will make your house too cold for comfort. There are ways to "bottle" this excess daytime heat so that your house will be comfortable all day *and* night.

The south side of this house will contain plenty of glass to admit sunlight. A thick concrete slab floor and water bottles on shelves will store energy for use at night when insulated shutters will cover the glass.

The Pueblo Indians used thick adobe walls; the temperature within medieval castles with their thick stone walls, floors, and ceilings changed very little through the day. Jacques Michel and Felix Trombe at their laboratory in the French Pyrenees built a massive concrete wall just inside large south-facing windows. They covered the rough side of the wall nearest the windows with flat black paint to increase heat absorption and reduce reflection. Air between the wall and window is heated, rises, and passes out through slots into the room while cooler air enters through ducts at the base of the wall. Dampers at the top of the wall are closed at night to prevent reverse thermosiphoning, and cold air settles into a basin at the bottom of the Trombe wall. Massive walls and floors absorb the heat and store it as their temperatures slowly rise.

The effectiveness of passive solar architecture can be enhanced by burying much of the house. The natural 55-degree temperature of the soil will reduce heat loss in the winter and modify heat gains in summer. Insulated shutters or covers for the large south windows can be closed at night or on cloudy days. If the inside of each cover is coated with aluminum foil, additional sunlight can be reflected into your home in daytime when the cover is lowered onto the ground.

If walls and floors are to absorb, store, and radiate heat, the insulation must be *outside* the walls not within them as is customary in wood frame buildings. This practice is unfamiliar to many builders, and it does create some problems. The insulation must be protected from moisture, light, and animals that might destroy it. Rigid styrofoam or polystyrene board insulation can be bonded to concrete walls by placing it in the framework before pouring the concrete. Fiberglass-reinforced concrete or moisture-resistant plywood can be used to cover the insulation.

Some passive solar homes use stacks of water-filled 55-gallon drums or open shelves filled with jugs of water to absorb heat instead of a Trombe wall. At latitudes below 35 degrees where the winter sun does not hover near the horizon, a roof pond can absorb solar energy during the day and radiate heat into the living space below at night. Insulating panels above the shallow pond slide into place at night to reduce heat losses to the cold air. In summer, the pond can be covered at sunrise and opened at night to radiate heat absorbed from the house during the day.

Passive solar house with Trombe wall and insulated reflective shutter.

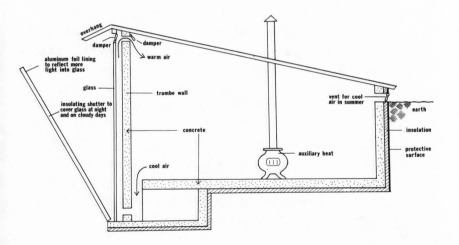

TRAPPING SOLAR HEAT

To retain the heat absorbed by concrete, water, bricks, or stone, passive solar homes must be very well insulated, caulked, and weather-stripped. Except for the south side, there should be few windows, and those should be double- or triple-glazed and covered at night with insulated shutters or shades. Trees or earth can screen the house from the wind and summer sunlight.

ELECTRICITY FROM SUNLIGHT

Why not convert solar energy to electrical energy and eliminate the need for coal, oil, or nuclear energy to generate electricity? By adding boron and phosphorus to thin silicon wafers, it is possible to produce photovoltaic cells (solar batteries) that generate electricity when illuminated. These amazing electronic devices are being used in photographic exposure meters, light meters, and as power sources in satellites and in a few remote areas of the world. But extensive use of these quiet, reliable, nonpolluting cells is many years away because they are very expensive. Electricity from sunlight requires an investment of $15,000 per kilowatt—40 times the cost of conventional power sources.

In addition, there are three other major problems in making electricity from sunlight: (1) the sun doesn't always shine, (2) the atmosphere absorbs a good portion of the energy, and (3) the sunniest areas on earth are usually not near major population centers.

One way to overcome these problems is to build solar power plants in space. Tests have shown that microwaves can be used to transmit electric power. Two recent developments in technology have made this possible: the amplitron, which can generate microwave energy, and the Schottky-barrier diode, which converts microwaves to direct current electricity.

Microwave generators are now mass-produced for use in microwave ovens. A single amplitron can produce 5,000 watts of microwave power at 90 percent efficiency, and with wave guides the microwaves can be transmitted in nearly parallel beams.

Because clouds are virtually transparent to microwaves, the energy from solar power plants can be transmitted through our atmosphere with minimal absorption. This fact has led Peter Glaser to suggest that solar power stations be built in space. These stations could beam microwaves to receiving stations on earth where the microwaves would be converted to electricity. Each space station would consist of two three-mile-long panels of solar cells and mirrors that would always face the sun. Between the panels would be the transmitting antenna, a dish-shaped device 3,000 feet in diameter that would always be turned toward a receiving station on earth. The amplitrons within the antenna would need no glass enclosures in the vacuum of space. Hollow aluminum wave guides would direct the microwaves toward earth.

Although Glaser's idea is attractive, it is fraught with problems. First, solar cells are very expensive. To meet that difficulty, Gordon Woodcock of the Boeing Company has a more conventional design. He would build a huge array of thin mirrors made by coating thin plastic with aluminum. The mirrors would focus beams of sunlight onto a closed sphere at the focal point of the mirrors. Helium gas, heated within the sphere, would serve as the fuel to drive conventional turbines. Generators connected to the turbines would send electricity to amplitrons within a transmitting antenna similar to the one suggested by Glaser. Second, the cost of building such stations on earth and ferrying them into space is prohibitive. So also is the expense of shipping the raw materials to a space station and assembling the power plant there. The only economical way to obtain electricity from space stations is to build them in space from raw materials obtained on the moon. But that would require lunar mining stations and the construction of huge space colonies where people could refine the lunar ore and manufacture the solar power stations—a program that may seem more feasible a century from now.

EXPERIMENTS

1. SUN AND SHADE

Here's an easy way to demonstrate that there's energy in sunlight. Place a thermometer on a sheet of cardboard. Shade the bulb with masking tape used to bind the thermometer to the cardboard. Place the cardboard sheet in a shady location. When the temperature is steady, place the sheet in the sunlight. What happens to the temperature?

2. Heating with a Lens (Magnifying Glass)

The energy in a beam of sunlight can be concentrated to a point with a convex lens (magnifying glass). Use such a lens to make a point of light on a piece of paper. How long does it take before the paper starts to smoke? (Be careful! Do this experiment in a place where there is no possibility of producing a fire.)

Now, use a lens with a smaller or larger diameter. How does the area of the lens affect the time it takes to make the paper smoke?

Another way to see how the lens area affects its heating capacity is to focus sunlight onto the bulb of a thermometer. How much can you make the temperature rise in ten seconds or some other appropriate time? Then repeat the experiment with a lens that has a larger or smaller area. (If you have only one lens, cover one side with black paper. A hole with half the area of the lens can be cut in the paper.)

How are the temperature changes related to the area of the lens used to bring the light energy to a point?

3. The Sun's Changing Path

Before building a solar house you need to know where the sun rises and sets and how high it ascends at different times of the year. You can find out by mapping the sun's path across a model sky. You'll need a clear plastic dome or a fine mesh kitchen strainer to represent the hemispherical sky (celestial hemisphere). Put the dome or strainer on a sheet of heavy cardboard. Mark the outline of its base with a pencil. Remove the dome or strainer and make an X at the center of the circle you have drawn. The X represents your position at the center of the celestial hemisphere. Put the dome or strainer back in its original position and tape it to the cardboard.

Place the dome outside on a level surface shortly after sunrise. To map the sun's path, place the tip of a marking pen on the dome so that the shadow of the tip falls on the X at the center of the dome as shown in the drawing. Mark the point on the dome with your pen. The pen mark on the dome represents the position of the sun in the sky because it is directly in line with the real sun and the X that represents you.

Make marks like this at half-hour intervals throughout the day to obtain a map of the sun's path across the sky.

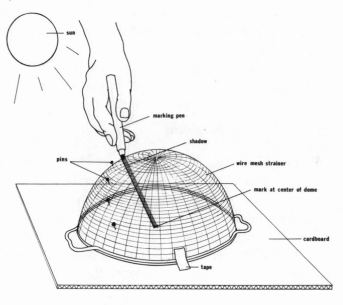

If you use a strainer, round-headed pins can be used to cast shadows onto the X mark. Leave the pins in the strainer. Together they will provide a colorful map of the sun's path when you run a piece of colored yarn through the pin positions as you remove them. In this way you can make permanent records of the sun's paths.

Repeat this experiment at different times of the year. Try especially hard to map the sun on or about the twentieth of September, December, March, and June. These dates will give you particularly interesting maps. Different colored marking pens (if you use a dome) or yarn (if you use a strainer) can be used for different dates. This will help you to see changes in the sun's path.

The sun maps in the photo were made at different times of the year. Which one was made in June? In December? The sun's path in March and September is the same. Which map was made in one of those months?

4. WHICH WAY IS SOUTH?

To build a solar home you need windows that face south. But which way is south? Don't use a magnetic compass; the compass needle points 20 degrees west of true north in northern Maine and 20 degrees east of north in the Idaho panhandle.

The simplest method is to find the North Star and

align it with several sticks to get a north-south line, but it may be difficult to do at night.

Here's an easy way. Drive a straight stick vertically into the ground and mark the tip of the stick's shadow throughout the day. Mark it very frequently around mid-day (11 A.M. to 2 P.M.). The sun will be highest in the sky, and due south, when the stick's shadow is shortest. (Not necessarily at noon. Clock time is not sun time.)

A line from the stick to the tip of its shortest shadow will point due south.

5. COLOR AND SOLAR ENERGY

Does the color of your house or roof have any effect on the solar energy absorbed or reflected? To find out, cover the bulbs of several thermometers with equal-size sheets (about 2 in. × 4 in.) of colored construction paper. Wrap the colored paper around each thermometer in the same way. A paper clip will hold the paper in place.

Note the temperature within each colored paper before placing the thermometers side by side on an insulating sheet of cardboard in a bright sunny place. Record the temperature within each colored sheet at one-minute intervals over a period of half an hour.

Which colored paper seems to be the best heat absorber? Which one seems to be the best heat reflector?

6. HEATING AIR WITH SOLAR ENERGY

Paint one of two identical aluminum pie pans with flat black paint. After the paint is dry, fix a thermometer to each pan's edge with masking tape. Be sure the tape covers the thermometer's bulb. Place each pan and attached thermometer inside a plastic bag and close the bags with tie bands. Place both pans in a cardboard box positioned so that the sunlight is perpendicular to the surface of the pans.

Check the air temperatures in each bag periodically. Which pan converts more solar energy to heat?

Repeat the experiment using a large and a small pan, both painted black. Which pan absorbs more heat?

7. HEATING WATER WITH SOLAR ENERGY

Paint one of two identical aluminum pie pans black as in Experiment 6. Put the same volume (50 or 100 ml) of cold water in each pan. Place both pans on an insulating sheet of cardboard in a warm, sunny place. Which pan seems to convert more solar energy to heat? How can you tell?

Repeat the experiment with two *black* pans—one large, one small. Which pan do you think will produce more heat?

8. THE SOLAR CONSTANT

Satellite measurements show that the sun delivers 1.95 cal./sq. cm./min. (1,353 watts/sq. m./min. or 7.15 Btu/sq. ft./min.) to the upper atmosphere, but how much gets to the earth's surface? To obtain a rough estimate, discard the upper half of a styrofoam cup and place the lower half in a small box. Surround the cup with insulating material.

Add black ink to cold water in a clear vessel until you

cannot see through the liquid. Pour 50 milliliters of the black water into the insulated cup and measure the temperature of the liquid. Place the box in sunlight when the sun is high in the sky. After a known time of a few minutes, record the temperature of the liquid again. How much heat, in calories, has the sun delivered to the water?

From the diameter of the cup you can determine the surface area of the water.

Use your measurements to estimate what fraction of the 1.95 cal./sq. cm./min. striking our atmosphere reaches the ground near where you live.

9. A MODEL PASSIVE SOLAR HOME

Cut a "picture window" in one side of a topless cardboard box. Tape a sheet of clear plastic wrap over the "window." Cut a small flaplike "door" on the opposite side of the box so you can place a thermometer inside.

Tape the box, top-side-down, to a sheet of cardboard that will serve as the floor of the "house." Place the house in a sunny place with the picture window facing north. After a few minutes record the temperature inside the box, and turn the box around so that the window faces the sun directly. (Be sure the thermometer is shaded in some way.) What happens to the temperature inside the house?

10. STORING SOLAR ENERGY

Passive solar heating requires materials that can absorb and store heat for times when there is no sun.

To test a number of materials for their potential heat-storing capacity, fill small cans (6-ounce frozen juice cans work well) about halfway with different substances such as water, gravel, sand, dirt, salt, sawdust (pencil sharpener waste), lead shot, and paper. Place the cans in a 120-degree oven until they're all at the same temperature. (Be sure to cover the water so it can't evaporate.)

Remove the cans and record their temperatures periodically as they cool. Which material cools fastest? Slowest? Which would be best for storing solar energy?

11. A MODEL FLAT PLATE COLLECTOR

Cut the flaps from a cardboard box (about 8 in. × 12 in. × 4 in. deep) and fill the lower half with crumpled newspaper. Punch a hole in one corner of the box about 1½ inches from the top and one side. Push the spout of a funnel through the hole (see drawing on the next page). The funnel should rest in a corner of the box to collect water.

Cut a sheet of cardboard the same width as the box. Its length should be equal to the distance from the top of the funnel to the other end of the box. The cardboard should fit tightly to help hold the funnel in place. Cut a piece of heavy duty aluminum foil about two inches wider than the box and equal to the length of the box. Lay the aluminum on the cardboard sheet; fold the "extra" aluminum up onto the sides of the box and tape it firmly. Turn up the bottom edge of the aluminum foil to serve as a trough to carry water into the funnel. Punch a hole through the foil above the funnel so that water can run out of the box into a collecting cup.

Use a thick pin to punch six or seven holes in a straight line along one side of a giant ($\frac{5}{16}$ -in. diameter) plastic soda

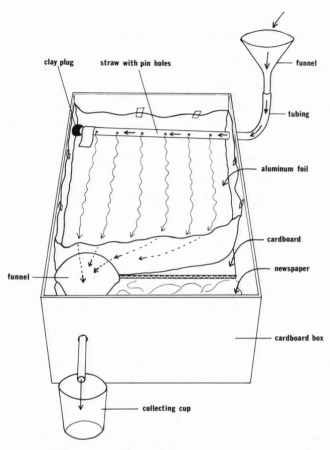

clay plug

straw with pin holes

funnel

tubing

aluminum foil

cardboard

newspaper

funnel

cardboard box

collecting cup

straw. Punch a hole in the upper corner of the box and push the straw through the hole. The straw should rest on the aluminum foil with the holes against the foil's surface. A clay plug over the end of the straw will force water to flow out the small holes. Use tape to hold the end of the straw firmly against the aluminum foil. The other end of the straw protrudes through the box and is connected to tubing leading from another funnel.

Let sunlight fall directly onto the aluminum foil. Measure the temperature of 50 milliliters of cold water and pour it into the upper funnel. The water will flow into the straw, out the tiny holes, and down the warm aluminum foil to the trough where it collects, flows into the funnel, and out the solar collector into a cup. You may have to jiggle the upper funnel to get all the water out of the straw.

In a real solar collector a pump would return the water to the top of the panel after it had passed through a storage tank. In this model collector you will have to remove the small cup and pour the water back into the upper funnel. Be sure to place another cup under the collecting funnel before you pour again.

After the water has run down the aluminum foil a few times, measure the water temperature again. How much has the temperature risen?

Quickly dry the aluminum foil and spray it with a quick-drying flat black paint. Place the box in the sun to speed the paint's drying. As soon as it is dry, repeat the experiment, using another 50 milliliters of cold water. How do you account for the larger rise in temperature this time?

Now cover the open side of the box with plastic wrap. Tape the transparent cover to the box and repeat the experiment again. How do you explain the larger temperature change this time?

12. BUILD A SOLAR HOT-DOG COOKER

You can use solar energy for cooking. Of course, you might be pretty hungry after a few cloudy days, but for sunny days try this solar hot-dog cooker.

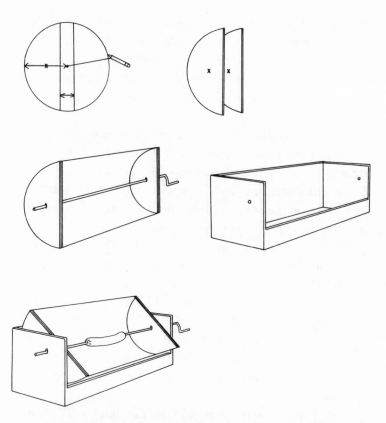

Draw two 18-inch circles on cardboard as shown. Make an X at the point halfway from the circle's center to its circumference. This X marks the focal line of the cooker. Cut each circle along lines comparable to the vertical lines on the circle in the drawing. Glue the two pieces from each circle together to form the ends of the cooker.

Tape a sheet of flexible light cardboard to the curved edges of the two end pieces. The total length of this concave "oven" should be a little longer than a hot dog. Cover the inside of the oven with aluminum foil. Use rubber cement to fasten the foil to the cardboard.

Punch holes in a box that will support your cooker, and run an unpainted coat hanger through the box and the X marks at each end of the oven. The sun's rays should focus on the coat hanger when the box and oven are properly aligned with the sun.

You're now ready to cook hot dogs—if the sun is shining!

13. BUILD A WINDOW BOX HEATER

For relatively little money you can build a solar heater that will fit into the south window of a room (see drawing).

First build an outer shell that will fit tightly inside a south window. To increase its heating capacity, you can make the heater wider than the window if you wish. Use

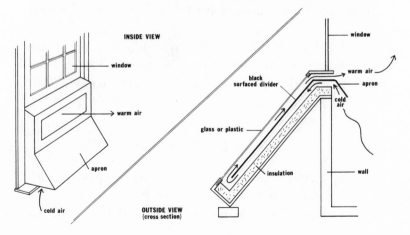

1- × 10-inch boards for the three sides and ⅜-inch exterior plywood for the bottom. Build a smaller inner shell from ¼-inch plywood. Insulation, 3½ inches thick, should lie between the two shells. Scrap wood spacers can be used to hold the shells at the proper separation.

A sheet of ¼-inch interior plywood is used to divide the inner shell into top and bottom. Leave a four-inch space across the bottom of the inner box so that cold air flowing down under the divider can ascend above the divider as it warms. Furring strips are used to support the divider about two inches below the top of the inner shell. Nail the divider in place and paint it with flat black paint.

Build a connecting "tunnel" between the box and the window using 1- × 10-inch boards. It too should have a divider with an apron (see drawing) where it enters the room so that only cool air near the floor enters the heater.

Use one-inch pine boards to cover the insulation between the two shells. Lay a sheet of clear plastic (Sun-Lite is good) over the box. Nail or staple wood strips to the edges of the plastic to hold it firmly against the pine boards.

If old storm windows are available, you could use them to make a transparent cover for the solar heater.

Be sure the heater fits tightly into the window. Fill any openings around the frame with insulation.

SOLAR ENERGY QUIZ

1. Does the sun always rise in the same place (east) and set in the same place (west)?
2. Does the sun's midday altitude change with the seasons?

3. During which season will the sun shine most directly on the south side of a house?
4. Can sunlight ever enter the north windows of a house?
5. What color clothing is best to wear on a cold, sunny winter day?
6. A black garbage bag and a white garbage bag are filled with air and left in a sunny place. In which bag will the air be warmer?
7. Must a passive solar home face due south?
8. Name a material often used to store heat in a solar heated home.
9. Would it make sense for a family to use solar cells to generate their own electricity?
10. Should insulation always be placed *within* the walls of a house?

CHAPTER

CHECKING
FOR ENERGY LOSERS AND SAVERS
AT HOME AND SCHOOL

THE CHECKLISTS and audit technique in this chapter will help you, your family, and your school reduce energy use and, thereby, costs.

ENERGY CHECKLISTS

Use checklists to make a thorough analysis of the use of energy in your home or school. Some checklists were provided earlier in this book. You may find additional items to add to the checklists as you make surveys.

APPLIANCES

Use the checklist in Chapter 3, pages 58–61.

HEATING AND COOLING

Use checklists on pages 65, 91, 125–127, and the following:
- Are thermostats turned back at night? At other times?
- Are thermostats set at 78 degrees or higher in summer?
- Is the heat in unoccupied rooms turned off?
- Would an attic fan cool your house?
- Does your fireplace have glass doors so that it can be closed off after use to prevent warm air from escaping through the chimney?
- Are air-conditioned rooms closed off?
- Are windows rather than air conditioners used for cooling on cool days?

INSULATION (SEE CHAPTER 4)

- Is the attic insulated? Ventilated?
- Are the walls insulated? (You can probably tell by removing the plate that covers an electric outlet on an exterior wall. (*Be sure to turn off the main switch before you do this.*)
- Are floors over cold areas insulated?
- Are floors over cold areas carpeted?
- Is the access door to the attic insulated?

TRANSPORTATION

Use checklist on pages 117–119.

WINDOWS AND DOORS

- Are there storm doors and storm windows?
- Are shades or draperies, preferably insulated ones, drawn at night and on cold, cloudy days?
- Do they fit tightly over windows and glass doors?
- Are windows used to provide passive solar heat in winter?
- Are shades or draperies drawn in summer to reduce solar heating?
- Are windows and doors caulked and weather-stripped?
- Do doors and windows close tightly?
- Does your draftometer indicate any air leaks?
- Are windows used for cross ventilation in summer?
- Are there vestibules around entrances?

AROUND THE HOUSE AND SCHOOL

- Do plants and trees provide a windbreak?
- Are there trees and a roof overhang to shade buildings from the summer sun?
- Is there a clothesline outside?
- Are electrical tools clean and lubricated?
- Are drills and saws sharp?

AN ENERGY AUDIT

In Chapter 4 you learned how to determine heat losses due to infiltration and conduction. You can use those techniques to calculate the heating losses from any building, or you can use an easier method developed by *Project Retrotech.*

In the Retrotech method, infiltration losses are determined by using a draft index. The index is simply the number of air turnovers per hour. It is calculated in the way described in Chapter 4, page 89.

Seasonal heat losses are determined not from degree-days × 24 hours, but from heating units and district heating factors. One heating unit (HU) is 100,000 Btu's; that is, the *useful* heating energy in a gallon of fuel oil, 15 pounds of coal, 120 cubic feet of natural gas, or 30 kilowatt-hours of electricity. The district heating factor for your area is the number of seasonal degree-days for your town or city divided by 4,000. (A heating factor of 1 is equal to 4,000 degree-days.) Since 24 × 4,000 is very nearly 100,000, the Retrotech method substitutes the district heating factor (DHF) for the number of degree-days × 24 hours to obtain heat losses in heating units rather than Btu's. You can easily convert heating units to Btu's. Simply multiply heating units by 100,000.

The pages that follow suggest a format that you can use to find heat losses for a building using the Retrotech method. To help you see how the method works, sample data for a house are presented.

First, you should list the structural components of the house so that R values can be determined from Table IV in Chapter 4. For the sample house used to illustrate, we have the following structure:

Dimensions—32 × 38 ft., one story, 8-ft. ceilings.
8 windows—all 32 × 48 in., single-glazed, double-hung, wood, good fit, no storm windows, but well weather-stripped.

2 exterior doors—2 ft. 8 in. × 6 ft. 8 in., solid wood, tight fit, no storm doors, but weather-stripped.

Walls—2 × 4 in. frame, ⅜-in. gypsum board inside, no insulation, ½-in. plywood, building paper, clapboard siding.

Ceilings—½-in. gypsum board, no insulation.

Roof—½-in. plywood, building paper, asphalt shingles.

Floors—hardwood flooring on ½-in. plywood throughout.

Crawl space—enclosed, average tightness, no trapdoor.

Heat source—fuel oil.

Degree-days—6,000.

Second, use the list of materials and R value tables to make a chart like the one below.

SURFACE	COMPONENT (R) + COMPONENT (R) + ...	R — TOTAL
Glass	Single-glazed, 1.0	1.0
Doors	Wood, 2.0	2.0
Walls	Air, 0.68; gypsum, 0.32; air, 0.9; plywood, 0.63; paper, 0.06; clapboard, 0.81; air, 0.17	3.6
Ceiling	Air, 0.68; gypsum, 0.43; air, 0.68	1.8
Floor	Air, 0.68; hard wood, 0.71; plywood, 0.63; air, 0.68.	2.7

Using the data given and Table IV in Chapter 4, we find that the draft index for the sample house is 1.5.

Use a chart like the one opposite to assemble figures you will need to calculate heat losses. The data shown are for the sample house.

SURFACE	DIMENSIONS (FT.)	AREA (SQ. FT.)	TOTAL AREA (SQ. FT.)	R VALUE
Glass	2.7 × 4	10.8	86.4	1.0
Doors	2.7 × 6.7	18	36	2.0
Walls	32 × 8, 36 × 8	(512 + 576) − 122*	966	3.6
Ceiling	32 × 36	1,152	1,152	1.8
Floor**	32 × 36	1,152	1,152	2.7

* Subtract window and door areas from wall area.
** The Retrotech method uses a floor exposure factor (FEF) to take into account variations in houses.

STRUCTURE	FEF
House on posts, floor exposed underneath	1.0
Crawl space skirted, or rock wall basement, or two feet of basement above grade	0.8
Slab, or tight crawl space, or tight basement	0.5

With this information, you can calculate heat losses.

Infiltration losses:

floor area $\times$ height = volume of air heated
$$1{,}152 \times 8 = 9{,}216 \text{ cu. ft.}$$
volume of air $\times$ draft index $\times$ DHF $\times$ specific heat = heating units
$$9{,}216 \times 1.5 \times 1.5^* \times 0.02^{**} = 415 \text{ HU}$$

$* \ \text{HF} = \dfrac{6{,}000 \text{ DD}}{4{,}000 \text{ DD}} = 1.5$

$** \ 0.018 \ \dfrac{\text{Btu}}{\text{cu. ft. F}}$ is approximately the 0.02 used by Retrotech.

Potential savings:

If the draft index were reduced to 1.0 by adding storm doors and windows, the heat required would be

$$9{,}216 \times 1.0 \times 1.5 \times 0.02 = 276 \text{ HU}$$

This would save $415 - 276 = 139$ HU, or $139 \times 100{,}000 = 13.9$ M Btu's per season.

Conduction losses through windows:

$$\frac{\text{glass area} \times \text{DHF}}{\text{R value}} = \text{heating units}$$

$$\frac{86.4 \times 1.5}{1.0} = 130 \text{ HU}$$

Potential savings:

Adding storm windows would increase R to 2.0

$$\frac{86.4 \times 1.5}{2.0} = 65 \text{ HU}$$

This would save $130 - 65 = 65$ HU, or 6.5 M Btu's per season.

Conduction loses through doors:

$$\frac{\text{area} \times \text{DHF}}{\text{R value}} = \text{heating units}$$

$$\frac{36 \times 1.5}{2} = 27 \text{ HU}$$

Potential savings:
Adding storm doors would increase R to 3.

$$\frac{36 \times 1.5}{3} = 18 \text{ HU}$$

This measure would save $27 - 18 = 9$ HU, or 900,000 Btu's per season.

Conduction losses through walls:

$$\frac{\text{wall area} \times \text{DHF}}{\text{R value}} = \text{heating units}$$

$$\frac{966 \times 1.5}{3.6} = 403 \text{ HU}$$

Potential savings:
Adding 3½ inches of blown-in cellulose insulation to walls would increase R by 13, giving R value of 16.6. The heat losses would then be

$$\frac{966 \times 1.5}{16.6} = 87 \text{ HU}$$

This measure would save $403 - 87 = 316$ HU, or 31.6 M Btu's per season.

Conduction losses through ceiling:

$$\frac{\text{ceiling area} \times \text{DHF}}{\text{R value}} = \text{heating units}$$

$$\frac{1{,}152 \times 1.5}{1.8} = 960 \text{ HU}$$

Potential savings:

Adding 12 inches of batt insulation in attic would increase R from 1.8 to 39.8. The heat losses would then be

$$\frac{1{,}152 \times 1.5}{39.8} = 43.4 \text{ HU}$$

This measure would save 960 − 43.4 = 917 HU, or 91.7 M Btu's per season.

Adding six inches of insulation would increase R from 19 to 20.8. Heat losses would then be

$$\frac{1{,}152 \times 1.5}{20.8} = 83.1 \text{ HU}$$

This would save 960 − 83.1 = 877 HU, or 87.7 M Btu's per season.

How much heat is lost through the floor? What heat savings would be realized if 3½ inches of batt insulation were installed below the floor?

POTENTIAL SAVINGS

To translate potential savings in Btu's into dollars, you need to know the cost of carrying out an energy-saving suggestion. For instance, if fuel oil sells for $1.00 per gallon, adding storm doors might not be a good investment. A saving of 900,000 Btu's per year in heat losses translates to nine gallons of oil, or $9.00 at $1.00 per gallon. It would take twenty-two years to return the investment on two $100 storm doors.

Insulating the ceiling with six inches of mineral wool batts at 25¢ per square foot would cost $288, but it would save 87.7 M Btu's—that's 877 gallons of oil. At a dollar per gallon, the investment would pay for itself in four months. If the ceiling already had six inches of insulation, it might not be worthwhile to invest $288 to save an additional 44 HU ($44 in oil costs) because it would take six and a half years to return the investment.

Only after making an energy audit can you make reasonably intelligent decisions about the financial advisability of carrying out various conservation measures.

9

OUR ENERGY FUTURE

By NOW you must realize that the energy crisis has no simple, inexpensive solution. Oil production in this country peaked a decade ago, giving rise to ever-increasing imports of this energy source. To avoid a national economic crisis we must both reduce our demand for oil and find suitable substitutes.

There are several ways we may proceed. One is to continue to import OPEC oil, whatever the cost, to meet our energy demands—a path than can lead only to continued devaluation of the dollar, an unfavorable balance of trade, and economic disaster.

We can reduce our demand for oil by following conservation measures such as those described in this book. As a nation of individuals, we have made a good start toward conserving energy, but we still need electricity, even if we use less; we still need gasoline, fuel oil, and natural gas, even if we use less. We need to find ways to replace oil as an energy source.

Solar energy certainly offers individual families a way to reduce the need for oil in generating hot water and space heat, but the present cost of photovoltaic cells makes the conversion of sunlight to electricity a very expensive process—one that American consumers cannot afford.

Fusion power would be a marvelous solution, but we have no proof that it can ever be made to work. Research in this field should continue because the potential is vast; however, it certainly offers no immediate solution to the energy crisis.

Ocean thermal power is another possibility that may serve as a partial solution in the future, but it too provides no immediate alternative to oil.

Oil shales contain vast amounts of the liquid gold, but environmental and energy costs involved in squeezing oil from marl may prove too high.

The production of uranium, like that of oil, has peaked in this country, and we have no way of increasing world or national uranium supplies; yet, other nations with oil deficiencies have boldly decided to use nuclear energy as an alternative to oil. France plans to produce 35 percent of its total energy from nuclear sources by the turn of the century. To fuel these nuclear reactors, France has already built, and is continuing to build, breeder reactors—reactors that produce not only power but more fissionable fuel than they consume. The fuel generated is plutonium, which can be separated from waste products and returned to the reactors as a power source. Unfortunately, plutonium can also be used to make atomic bombs, and the United States government has been reluctant to build breeder reactors for fear that plutonium, should it fall into the hands of terrorists, might be used to create an international nightmare.

Nuclear reactors in this country have had an enviable safety record. No one has been killed in a nuclear accident in the last twenty-five years. During the same period 6,500 were killed in coal mines. Despite the record, the accident at Three Mile Island has soured the public and Congress on the future development of nuclear power of any kind.

Coal, our most abundant energy source, accounts for 25 percent of our energy supply. By 1985 our current annual production of 600 million tons could be doubled. Utility companies are already being forced to convert from oil to coal. Clearly, coal, if developed properly, can reduce our dependence on foreign oil. But coal is not without problems. Coal mining is a dangerous occupation, and strip mining has had disastrous environmental effects. The wastes from burning coal are dirty and pollute the atmosphere. Further, coal is difficult to burn and expensive to transport.

Nevertheless, it is possible to improve mining safety, control the ill effects of strip mining, reduce the polluting effects of coal combustion, and convert solid coal to gas or liquid. All of these measures will cost money, and the price of coal will rise, but imported oil is well beyond what we as a nation can afford, and money spent on improving coal as a fuel and as an energy source is money spent within this nation.

The conversion of coal to gas is an old technique. Before electric lights were common, many cities were illuminated by coal gas. With the advent of electricity and cheap natural gas, coal gas was no longer competitive as a light or fuel source. Today, as natural gas supplies dwindle, coal gas at $4.00 per million Btu's is competitive with oil at $20 per barrel—a price OPEC surpassed in 1978; yet, we are not producing coal gas. We need to and soon!

During World War II Germany used the Fischer-Tropsch process to convert coal to oil. Today the only plant using this process is in South Africa, where limited oil and abundant coal made the process economically feasible twenty years ago.

With modern chemical catalysts the conversion of coal to gas—or to liquid, where necessary for transportation and ease of combustion—provides an energy source that would be economically competitive with oil.

The possible increase in average world temperature, due to the global greenhouse effect created by growing levels of carbon dioxide in the atmosphere, is *not* a problem that will be solved by using coal to replace oil. Only conservation measures can reduce that effect today.

However, the increased use of coal, coupled with energy conservation measures on both individual and national levels, can have immediate effects in reducing our dependence on foreign oil. But we must begin *now*.

Once we have learned to substitute coal for oil in a conservation-minded society, we must persist in our search for long range solutions to producing power on a large scale, whether it be fusion power or a combination of ocean thermal, geothermal, hydro, nuclear, and perhaps, in the next century, solar power from satellite stations.

ANSWERS
TO QUESTIONS

CHAPTER 3

1. 330 kwh, $16.50.   2. 960 kwh, $48.   3. 1,400 kwh, 35 percent.   4. 120 kwh, $6.   5. 800 kwh.   6. $20.   7. 400 kwh, $20.   8. 600 kwh, $30.   9. No.   10. No. No. 11. 1,200 W.   12. 0.2 kwh, 720,000 J.   13. 1¢.

CHAPTER 4

Quiz A: 1. January.   2. September.   3. February has only 28 days.   4. $130.   5. Honolulu.   6. Fairbanks.   7. Buffalo will consume twice as much.   8. The average daily temperature in Honolulu never goes below 65; it does in Miami.   9. 300 DD, 33 percent.   10. 1 Btu. *Quiz B:* 1. T.   2. T.   3. F.   4. T.   5. F.   6. T.   7. F. 8. T.   9. T.   10. F.

CHAPTER 5

Quiz A: 1. F.   2. T.   3. F.   4. T.   5. F.   6. F. 7. T.   8. F.   9. T.   10. F. *Quiz B:* 1. 31.2 mpg.

2. Mountain travel, cold weather, snow.   3. Express high-way travel.

CHAPTER 6
1. d.   2. b.   3. d.   4. a.   5. b.   6. c.   7. a.
8. c.   9. d.   10. a.

CHAPTER 7
1. No.   2. Yes.   3. Winter.   4. Yes, in the summer and late spring.   5. Dark clothing.   6. The black one.   7. No, a few degrees east or west of south is satisfactory.   8. Water, stones.   9. No, it's too expensive.   10. No.

BIBLIOGRAPHY

Alternative Energy Handbook. Emmaus, Pa.: Rodale Press, Inc., 1979.

Anderson, Bruce with Riordan, Michael. *The Solar Home Book.* Harrisville, N.H.: Brick House Publishing Company, Inc., 1976.

In the Bank or up the Chimney. Washington, D.C.: U.S. Government Printing Office, 1977.

Keyes, John. *Harnessing the Sun to Heat your Home.* Dobbs Ferry, N.Y.: Morgan & Morgan, Publishers, 1974.

Morrison, James W., Editor. *The Complete Energy-Saving Home Improvement Guide.* New York: Arco Publishing Co., Inc., 1978.

Phillips, Owen. *The Last Chance Energy Book.* New York: McGraw-Hill, 1979.

Residential Energy Conservation. Washington, D.C.: U.S. Government Printing Office.

Tips for Energy Savers. Washington, D.C.: Federal Energy Administration, 1977.

Walker, Harry O. *Energy: Options and Issues.* Davis, Calif.: University of California, 1977.

Wolfe, Ralph and Clegg, Peter. *Home Energy for the Eighties.* Charlotte, Vt: Garden Way Publishing, 1979.

INDEX

ABOUT THE AUTHOR

R OBERT GARDNER is head of the science department at Salisbury School, Salisbury Connecticut, where he teaches physics, chemistry, and physical science. He did his undergraduate work at Wesleyan University and has graduate degrees from Trinity College and Wesleyan University. He has taught in a number of National Science Foundation teachers' institutes and is the author of several science books, including *Space: Frontier of the Future, This Is the Way It Works,* and *Moving Right Along* (with David Webster).

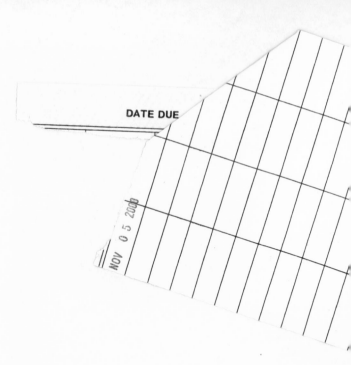